Daniel James Pedersen

The Eternal Covenant

Theologische Bibliothek Töpelmann

Herausgegeben von
Bruce McCormack, Friederike Nüssel
und Christoph Schwöbel

Band 181

Daniel James Pedersen

The Eternal Covenant

Schleiermacher on God and Natural Science

DE GRUYTER

ISBN 978-3-11-054080-2
e-ISBN (PDF) 978-3-11-054230-1
e-ISBN (EPUB) 978-3-11-054128-1
ISSN 0563-4288

Library of Congress Cataloging-in-Publication Data
A CIP catalog record for this book has been applied for at the Library of Congress.

Bibliografische Information der Deutschen Nationalbibliothek
Die Deutsche Nationalbibliothek verzeichnet diese Publikation in der Deutschen Nationalbibliografie; detaillierte bibliografische Daten sind im Internet über http://dnb.dnb.de abrufbar.

Druck und Bindung: CPI books GmbH, Leck
♾ Gedruckt auf säurefreiem Papier
Printed in Germany

www.degruyter.com

For my friends

Acknowledgments

The advice, work, and support of many people contributed to this book. I am so grateful to all of them. I especially want to thank the following friends and colleagues who often went above and beyond to encourage, criticize, and improve this work: Christopher Lilley, Derek Woodard-Lehman, David Chao, Jeffrey Skaff, Bruce McCormack, John Bowlin, Wentzel van Huyssteen, Daniel Garber, and Robert Adams. I am deeply indebted to each of them, as well as to the many people who contributed in smaller ways who remain unnamed.

Most of all, I would like to thank my wife, Danielle, without whose support, encouragement, kindness, and patience this book would simply not have been possible.

Contents

Abbreviations

GL Friedrich Schleiermacher, *Der christliche Glaube nach den Grundsätzen der evangelischen Kirche im Zusammenhange dargestellt.* Second edition. Edited by Rolf Schäfer. Berlin: Walter de Gruyter, 2008.

Chapter 1 Introduction: The Eternal Covenant

> "All men, Socrates, who have any degree of right feeling, at the beginning of every enterprise, whether small or great, call upon God. And we too, who are going to discourse on the nature of the universe, how created or how existing without creation, if we be not altogether out of our wits, must invoke the aid of Gods and Goddesses and pray that our words be acceptable to them and consistent with themselves."
>
> - Plato, *Timaeus*

These days it is fashionable to believe, especially in the face of the encroachments of modern science, that there is no freedom without indeterminacy, and no love without indeterminate freedom. This, it is supposed, is as true for God as it is for creatures.

It is remarkable how far this assumption extends. In contemporary conversations on theology-and-science, theorizing a way for God to act *within* the forces and powers of the natural world, without interrupting them, has become its own cottage industry.[1] That this is even a genuine problem is far from obvious, and yet the solutions keep coming. Almost inevitably they trade on an indeterminist reading of quantum theory, reified into a *really* indeterminate universe.[2] God, on this account, need not topple the structures of the universe because those struc-

---

1 See, for example: Alvin Plantinga, *Where the Conflict Really Lies* (Oxford: Oxford University Press, 2011); Robert John Russell, *Cosmology: From Alpha to Omega* (Minneapolis: Fortress, 2008); Thomas F. Torrance, *Divine and Contingent Order* (Edinburgh: T&T Clark, [1981] 1998); Keith Ward, *God, Chance, and Necessity* (Oxford: Oneworld, 1996). For a summary of the state of the field as of 2003, especially of the work done in connection with the Vatican Observatory and the Center for Theology and the Natural Sciences see Wesley Wildman, "The Divine Action Project, 1988–2003", *Theology and Science* 2, 1 (2004), 31–75. For a powerful critique of this project at large see Nicholas Saunders, *Divine Action and Modern Science*, (Cambridge: Cambridge University Press, 2002).

2 See Alvin Plantinga, *Where the Conflict Really Lies*, 92, n. 1.; Russell, *Cosmology*, 151; Ward, *God, Chance, and Necessity*, 134. See also , Robert John Russell, Nancy Murphy, and Arthur Peacocke, eds. *Chaos and Complexity: Scientific Perspectives on Divine Action* (Vatican City: Vatican Observatory Publications, and Berkeley: Center for Theology and the Natural Sciences, 2000); Robert John Russell, Philip Clayton, Kirk Wegter-McNelly, and John Polkinghorne, ed., *Quantum Mechanics: Scientific Perspectives on Divine Action* (Vatican City: Vatican Observatory Publications, and Berkeley: Center for Theology and the Natural Sciences, 2002).

DOI 10.1515/9783110542301-001

tures are already open. God has elbow room to work, unrestrained by God's own design. The effort to secure the universe as indeterminate is in part an effort to secure divine freedom of a certain kind, the freedom to act within or upon a world from which, it is assumed, divine activity would otherwise be absent.

The assumption that indeterminacy is a necessary condition of divine freedom is not limited to theology and science specialists. It also permeates much of last century's work on the doctrine of God. From philosophers of religion to scholars in the recent trinity and election debates, there is a running assumption that indeterminate divine action of the divine essence is as necessary to secure freedom within God's own life as indeterminacy in nature is to secure free divine interaction with the world.[3] Indeterminate divine action, it is assumed, is the only alternative to compulsion. It is indeterminate divine freedom that makes God's grace genuinely free, and so, genuinely grace, and it is this free grace, this grace which needn't have been, that secures the love of God as love. For, it is assumed, what you cannot decline to love, you cannot genuinely love. While some theologians worry about divine action being constrained by the world, other theologians worry about divine action being constrained by the divine nature itself.

For more than a century, many theologians and their interlocutors have taken the bait, but it was not always so. In a similar situation, and facing remarkably similar assumptions, Friedrich Schleiermacher gave an account of the Christian faith and its relation to natural science which required none of the indeterminacy which contemporary accounts of many kinds judge essential. On Schleiermacher's account, the "problem" of how to secure divine freedom through indeterminacy, in either the world or in the divine life itself, is a false problem. In fact, he argues the opposite: that the indeterminacy which both advocates and critics judge necessary for God is both incoherent and impious. Indeterminacy is instead an imperfection of the definite action of a definite God: a failure of divine power, divine wisdom, or both. And, in fact, not only are worries over *both* questions of divine action—divine intervention vis-à-vis the world, or indeterminacy vis-à-vis the divine nature—equally problematic, they are problematic in the same way because they are simply two facets of one and the same false problem. Schleiermacher's alternative account transcends both at once. And, as an account which explicitly proceeds from piety, it does not admit of dead forces, or blind necessity, and least of all the superfluity of the Supreme Being, but insists instead that criticism of this sort is not only compatible

---

3 See Brian Leftow, *God and Necessity* (Oxford: Oxford University Press, 2012); Michael T. Dempsey, ed. *Trinity and Election in Contemporary Theology* (Grand Rapids, MI: Eerdmans, 2011).

with piety, but necessitated by it. The following chapters unpack each of these claims in turn.

It might sound unbelievable to pair the seemingly disparate problems of such different interests together. It might sound even more unbelievable to promise one and the same solution to all. Yet this is exactly what Schleiermacher himself did when he proposed his "eternal covenant" between the Christian faith and natural science. What exactly that covenant amounts to, and how exactly it is supported, is the subject of this work. Schleiermacher's account, as we will see, is not only the consequence of method, or of obeisance paid to the cultural prestige of natural science, but the result of leveraging commitments shared by both the Christian faith and natural science. For his case, he summons shared principles and deploys philosophical and theological arguments, many of which have not been fully appreciated, and some of which have not even been recognized. The eternal covenant, is the result of this intricate effort.

The Eternal Covenant in the *Letters to Lücke*

In his open *Letters to Lücke*, Schleiermacher famously proposed an "eternal covenant between the living Christian faith, and completely free, independent, scientific inquiry, so that faith does not hinder science and science does not exclude faith."[4] In itself, this neither amounts to an argument, nor even a clear program. In consequence, a good deal of labor has been devoted to clarifying what Schleiermacher meant when he proposed this eternal covenant, and how what he meant solves the issues at hand—if it is a solution at all. While there are important insights in each of these interpretations, none are wholly adequate.

The existing work on the eternal covenant often resolves Schleiermacher's proposal into two generalized camps, most clearly explained by Andrew Dole as the "segregation model" and the "accommodation model."[5] The former, represented by Richard Brandt's reading,[6] secures the compatibility of theology and science by, as the type suggests, segregating them. Theology and science can never, in principle, conflict because they have nothing to do with each other.

4 Friedrich Schleiermacher, *On the Glaubenslehre: Two Letters to Dr. Lücke*, trans. James Duke and Francis Fiorenza (Atlanta: Scholars Press, 1981), 64, Emphasis added.

5 Andrew Dole, *Schleiermacher on Religion and the Natural Order* (Oxford: Oxford University Press, 2010), 140, 144.

6 Dole, *Schleiermacher on Religion and the Natural Order*, 140. See also Richard Brandt, *The Philosophy of Schleiermacher* (New York: Harper, 1941), 261–262.

On this reading, Schleiermacher's eternal covenant amounts to something like Stephen Jay Gould's proposal for "nonoverlapping magisteria," or NOMA.[7] It is a riff on the fact-value divide of positivism, where feeling, or religious experience, is substituted for aesthetic or moral value. The idea that "faith does not hinder science, and science does not exclude faith,"[8] suggests two fields having less to do with each other, not more: as Schleiermacher puts it a few lines later, "that [the Christian consciousness] remain free from entanglements with science."[9] The segregation model is, at least on the face of things, a natural interpretation of the eternal covenant.

There are, however, problems with the segregation model. As Dole points out, it fails to make sense of Schleiermacher's proposal in his intellectual context: "The segregation model misses this fact that Schleiermacher's call for an eternal covenant was directly addressed to an intensifying conflict between the liberalism of which he had become a well-known proponent and an explicitly protectivist and anti-intellectual neopietism."[10] Schleiermacher's proposal would, in the segregation model, support an identical position to his opponents'. Demarcation along these lines heightens, rather than resolves, conflict between disciplines. Theologians of all kinds would become those very barbarians who, "fence themselves in with weapons at hand to withstand the assaults of sound research [...]."[11] In the final analysis, such a reading is too ironic to be likely.

In place of the segregation model, Dole suggests his own interpretation: the accommodation model.[12] This model, "sees Schleiermacher, first, refusing to place any limitations on scientific investigation and, second, imposing upon the religious an obligation to understand themselves and their religion in terms compatible with the sciences—in a word, accommodating to the past and future deliverances of natural-scientific research."[13] This interpretation has obvious advantages over the segregation model: it has no trouble situating Schleiermacher's proposal within his context, and it is clearly an alternative to the view of his opponents. Further, it acknowledges the authority Schleiermacher gave to natural science. In practice, it amounts to Christianity ceding as much as it can to the claims of natural science while nevertheless maintaining "a ser-

7 Steven Jay Gould, "Nonoverlapping Magisteria", *Natural History*, 106, (1997), 16–22.
8 Schleiermacher, *Letters to Lücke*, 64.
9 Schleiermacher, *Letters to Lücke*, 64.
10 Dole, *Schleiermacher on Religion and the Natural Order*, 144.
11 Schleiermacher, *Letters to Lücke*, 60.
12 Dole, *Schleiermacher on Religion and the Natural Order*, 140.
13 Dole, *Schleiermacher on Religion and the Natural Order*, 144.

ies of claims *over and above* but not in conflict with either the actual or the anticipated deliverances of natural-scientific investigation."[14] Concrete support for this latter aspect of the accommodation model can be found in Schleiermacher's rejection of miracles, and even a beginning of creation itself, where he seems to put this program to practice.[15] On these grounds and others, the accommodation model is a more plausible interpretation of the eternal covenant than the segregation model.

Dole is not the only reader of Schleiermacher to uphold to the accommodation model, or something like it. Less-developed versions of Dole's reading are standard in recent scholarship. Julia Lamm, for instance, claims that, "in the *Glaubenslehre*, [Schleiermacher's] concern is to show there is no contradiction between science and religion. Science, he says, sets certain limits to what can be said in the dogmatic enterprise."[16] Likewise, Robert Sherman describes Schleiermacher's project as "dissecting, purging, and rejoining" some dogmatic claims, so that others might be "rooted in assumptions not antithetical to science [...]."[17] With these descriptions, a pattern begins to emerge. The covenant is, on each count, attempting to avoid contradiction. Such contradictions are avoided by abandoning the commitments which generate them. And, in the accommodation model, it is at least largely the responsibility of the Christian faith to resolve conflicts with natural science by dispensing with *its* less important beliefs, beliefs which have become burdensome to itself and repugnant to others. Against the segregation model, this is much more plausible; as above, Schleiermacher calls for something like this in the *Letters to Lücke*. And, insofar as this interpretation is able to make sense of the instances where Schleiermacher does abandon commitments in potential conflict with natural science, the accommodation model is the superior model.

However, in spite of the superiority of the accommodation model, some puzzles remain. *Both* the segregation model *and* the accommodation model are methodological or programmatic resolutions to the problem. That is, they purport to solve conflict between the Christian faith and natural science with a procedure, a blueprint, a plan of action. However, it is not obvious that what Schleiermacher meant was restricted to a methodological solution, and, even if it is what he meant, it is not clear that it would be a good idea.

---

14 Dole, *Schleiermacher on Religion and the Natural Order*, 146.

15 Schleiermacher, *Letters to Lücke*, 60–61.

16 Julia A. Lamm, *The Living God: Schleiermacher's Theological Appropriation of Spinoza* (University Park, PA: Pennsylvania State University Press, 1996), 132.

17 Robert Sherman, *The Shift to Modernity: Christ and the Doctrine of Creation in the Theologies of Schleiermacher and Barth* (New York: T&T Clark, 2005), 22.

Methodological readings of the eternal covenant might imply that the covenant itself is a methodological proposal. That means that not only is the covenant best summed up in a generalized procedure, but it was actually proposed by Schleiermacher just as such a procedure. If so, as a starting point, the eternal covenant would amount to rules for avoiding or resolving conflict. Further, at least some of the conflict mentioned by Schleiermacher, and nearly always the conflict assumed by his interpreters, is a conflict between different *authorities* more than concrete beliefs. The conflict to be avoided or resolved is between natural science and the Christian faith writ large. Particular beliefs are adjusted subsequent to the right ordering of authorities in the hierarchy. Moreover, those authorities are not individual, but collective: the authority of disciplines as disciplines, the authority of natural science, and the authority of the Christian faith as guilds. If so, the eternal covenant is primarily *political*, not conceptual. In other words, the motivation for such an eternal covenant is not first and foremost that the claims and reasons of natural science bear on traditional theological claims, challenging them sometimes to the point of those claims' obsolescence, but instead that natural science as a discipline should be ceded intellectual territory in advance of *whatever* it may claim. Since, in the *Letters to Lücke*, Schleiermacher is concerned with resolving exactly this kind of public controversy, the eternal covenant is surely at least partly political. And yet the interpretation of the eternal covenant as *primarily* a program for resolving political conflicts between disciplines is both implausible and unsatisfying.

It is not obvious that such a political solution would have been either desirable or persuasive in context. If Schleiermacher's eternal covenant was no more than a rule for disciplines to arbitrate authority, then it—the covenant itself—could have no warrant in higher claims. Parties would have to begin by agreeing that the covenant, as a scheme for carving up intellectual territory, was a common good. But on what basis could they agree to this? Any basis would have to include reasons held in common, and any agreement sought would have to be on the basis of prior agreement. But in the segregation model, neither natural science nor the Christian faith can share common reasons. And although in the accommodation model they can share reasons, they cannot share authority. If the two disciplines somehow began by agreeing to the covenant, then fine, they would have agreed. But that then would render the covenant superfluous, since the purpose of the covenant is to mediate differences. But, if faith and science did not agree, what then? In that case, there would be no common basis to convince one another. Since the eternal covenant was proposed in a context of conflict and disagreement, its results cannot have been assumed from the start. And since both 'barbarian' Christians and 'unbelieving' scientists were already actively trying to deny one another authority, mutual regard could not even be assumed. If the eternal covenant involved a political solution—which it surely did in part—it could not have been a starting point. Just as Dole rightly argues that the segregation model is implausible when considered in context, since Schleiermacher's theological (not to mention scientific) opponents *desired* segregation, so the accommodation model, *as a starting point,*[18] is implausible on the exact same grounds: should parties have already agreed, they would not have required persuading, and, if they did not agree, Schleiermacher's proposal could not have been persuasive.

The second reason that the eternal covenant is not *primarily* a methodological resolution to a political conflict between disciplines is that Schleiermacher claims that it is his "firm conviction that the basis for such a covenant was already established in the Reformation."[19] Depending on how much one emphasizes the *basis* of the covenant versus the covenant itself, this claim is open to various readings. None of these, however, is wholly consonant with a political interpretation of Schleiermacher's proposal. A major theme of the Reformation was, broadly, the adjudication of authority, but natural science—or, in the sixteenth and seventeenth centuries, "natural philosophy"—was only a peripheral topic. And, where issues of natural science did arise, such as in the Lutheran theologian Andreas Osiander's introduction to Copernicus' *De Revolutionibus Orbium Coelestium*,[20] the solution was closer to segregation, not accommodation.[21] For his part, Schleiermacher clearly judged that the Reformers themselves held a deficient account of creation derived from a naive reading of *Genesis*.[22] It was accounts just like these which, he warned, Christians would have to surrender.[23]

Further, not only was there no strong tradition of accommodating the deliverances of natural science in the Reformation era, but, even if there was, it could not be between disciplines *as* disciplines. Natural science as a distinct professional enterprise was a relatively new concept (and guild) in the early *nineteenth* century. There was no equivalent *discipline* in the sixteenth century. It is difficult

---

18 I say this specifically to avoid implying that Dole thinks the "accommodation" model *must* be read as Schleiermacher's actual procedure. Instead, it could simply be a useful summary of a result. This latter view is perfectly compatible with my claim, and, even supported by it.

19 Schleiermacher, *Letters to Lücke*, 64.

20 Andreas Osiander, *Ad lectorem de hypothesibus huius operis*, 1543.

21 John Dillenberger, *Protestant Thought and Natural Science* (Notre Dame: University of Notre Dame Press, 1960), 45 – 47.

22 Friedrich Schleiermacher, *GL* §40.1; *The Christian Faith*, ed. H.R. Mackintosh and J.S. Stewart, trans. D.M. Baillie, *et al.* (Berkeley: Apocryphile, [1928] 2011), 150.

23 Schleiermacher, *Letters to Lücke*, 60 – 61.

to see how the eternal covenant could possibly have had its basis in the Reformation if the substance of the covenant is a rule for governing the authority of disciplines since the Reformers did not recognize natural science as a discipline in the modern sense, and even hesitated to grant the natural philosophy of their own day authority over faith claims.

Third, none of the explanations of the eternal covenant on offer explain why the covenant is properly described as either "eternal" or even a "covenant" at all. "Covenant" and "eternity" are not terms traded lightly in the Christian tradition. Covenants are divine compacts, predicated on divine faithfulness. They are agreements of common purpose, divine initiations or renewals of friendship between God and God's people. Even as a metaphor or rhetorical device, it is difficult to see how any methodological reading of the eternal covenant amounts to anything like a shared pursuit of common goods predicated on God's faithfulness. Considering this, the segregation model is most implausible. Rather than a common endeavor, it is actually their complete divorce. More like Adam's Fall than divine accord, the segregation of the Christian faith from natural science (and vice versa) carves up creation into two domains: the first, ruled by the divine sovereign, and the second, an area which rejects divine rule and claims its own independent authority. What is science's to gain is faith's to lose, and vice versa. Such an arrangement could hardly be called a covenant. The accommodation model, on the other hand, at least bears the appearance of agreement. It requires the cooperation of the Christian faith and natural science. Yet, even in that case, it is not clear how or why the Christian faith's deference to the deliverances of natural science is a common good—unless the good of the agreement consists chiefly in the Christian faith's avoidance of embarrassment. And even if this very modest good was a good to be pursued in common, it is not obvious why such a good or its pursuit should be "eternal" in any sense of the word. It seems instead a highly contingent good: good only for some, in some times and some places, specifically those times and places where natural science and the Christian faith are arranged as distinct authorities, and where their relative prestige makes natural science the default victor should conflict arise. If it is a covenant, it is no more than an ephemeral one.

Finally, Schleiermacher says that he wrote *The Christian Faith* to accord with his proposal for an eternal covenant.[24] But Schleiermacher's account of the nature system and the corresponding reasons for the authority of natural science in *The Christian Faith* are not themselves merely methodological. Instead, Schleiermacher's account relies on arguments, arguments which are developed over

24 Schleiermacher, *Letters to Lücke*, 64.

hundreds of pages. Had the related doctrines of *The Christian Faith* simply been consequences of an arrangement for carving up disciplinary territory, accompanying arguments would be unnecessary. Schleiermacher could have skipped the reasons for his doctrines and just announced those doctrines as part and parcel with his covenant. The eternal covenant, and the theological specifics that go along with it, could simply be sold as a package deal. If the eternal covenant is only a methodological proposal, it seems that Schleiermacher wasted his words, and many of them. However, none of the arguments Schleiermacher gives in *The Christian Faith* even rely on the covenant as a premise. In fact, the covenant does not support Schleiermacher's doctrinal claims, but is rather *supported by them.* His account of divine action follows from his account of divine being. That account is, in turn, predicated on God actually being the almighty-loving *Whence* on which we are conscious of being absolutely dependent. None of these arguments follow *from* the eternal covenant or any interpretation of it. Not only does the covenant as a methodological program fail to explain why Schleiermacher should provide arguments at all, but the covenant can't even be included as a step in those arguments.

Those seeking to explain the eternal covenant as a methodological proposal for its own time and place face many difficulties. But those who see promise in his ideas and argue for the eternal covenant as a path forward in the present face more burdens still, *if* Schleiermacher's solution was primarily a program for negotiating the authorities of disciplines. On both the question of the demarcation of disciplines, and of disciplines' relative authorities, thinkers of the last century had much to say, especially philosophers of science.

The so-called "problem of demarcation"—what makes natural science to be what it is and distinct from non-science—was a central question of twentieth-century philosophy of science. And if Schleiermacher's proposal is, in fact, a methodological proposal, it would have to be a central question in any future adoption of the covenant. This is true in either the segregation or accommodation models, or any interpretation of the covenant where the covenant is first and foremost a rule for negotiating the relative authority of disciplines. Regardless of the rubric proposed, particular disputed claims would need to be sorted under the proper authority, and would belong to one authority and not another. The resolution of any given conflict would then proceed deductively by the application of the clear authority of a discipline as a principle. It is, of course, this surefire sorting which makes a procedure like this so appealing at first glance. For example, under the segregation model all parties, presumably, agree that certain sorts of claims, topics, questions, and problems fall under the auspices of one, and only one, discipline. In this case, all parties agree to segregate faith and science. Some questions fall under the one, some under the other, none under both. Those claims that properly fall under the auspices of faith are faith's to claim incontestably; those that fall under the auspices of nature science are natural science's to claim. To settle a dispute over any given claim, *c*, all that is needed is to sort it properly. If *c* properly belongs to natural science, the question is necessarily ceded to natural science. The pre-arranged authority of disciplines serve as major premises, the assignment of particular claims to one discipline or another as minor premises, and so the final authority over any claim follows necessarily. The same procedure would hold true with the accommodation model—or any other "model" for that matter. All that changes model by model is the scope of authorities and the details of their division, not the procedure in general.

This is where the demarcation problem enters in.[25] Although no longer considered the central problem in the philosophy of science it once was,[26] failure to demarcate science from non-science remains a problem for the continuing relevance of any methodological interpretation of Schleiermacher's eternal covenant. Rules for sorting claims require *both* a reliable way to distinguish claims *and* clear kinds to sort those claims into. The absence of clear demarcation makes this process messy at best. Some claims, of course, might be obviously theological, and others obviously natural-scientific, but others still will fall somewhere in between, or somewhere else altogether. Since the utility of the covenant as a starting point for negotiating disciplinary authorities would lie precisely in its ability to cleanly sort claims, the messiness of the process undermines that utility. This is not to say that claims couldn't be sorted at all. They could be. But it would be a highly contestable process, not the deductive sorting that clear disciplinary boundaries, and obvious claim assignments, advertise. Since a major aim of the reading of the covenant as a method is precisely to resolve such contests before they begin, the covenant does not and cannot achieve one of its main aims. Instead, the covenant so interpreted would merely shift questions of authority. If faith is obliged to cede authority to natural science over certain questions, then the real fight becomes which claims count as natural

---

25 The history and diversity of the problem is vast. For one position and the state of debate in the early twentieth century see: Karl Popper, *The Logic of Scientific Discovery* (London: Routledge, [1932, trans. 1959] 2002). Perhaps the most memorably titled work on the topic is Rudolph Carnap's essay "Ueberwindung der Metaphysik durch logische Analyse der Sprache" ['The Overthrow of Metaphysics through Logical Analysis of Language'] *Erkenntnis*, 2 (1932).

26 One of the most important criticisms of demarcation was W. V. Quine's "Two Dogmas of Empiricism", *The Philosophical Review* 60, 1 (1951), 20 – 43. For an important alternative to science/non-science demarcation see Larry Laudan, *Progress and Its Problems* (Berkeley: University of California Press, 1977).

science, how to distinguish these, and so on. Contestation is by no means avoided.

Not only do readings of the eternal covenant as a starting point inadequately explain what Schleiermacher was doing in his own context, but the viability of such readings as normative proposals is problematic. They can't serve as the clear starting points that they appear to be. Any application is messy—so messy, in fact, that the covenant would serve little to no use in concrete application when actually adjudicating claims.

So, are interpretations of the eternal covenant as a general program wholly mistaken? Certainly not. In fact, this work will support something like the consequences Dole and others have drawn. However, whereas many interpreters have sought to couch the eternal covenant as a *starting point* for Schleiermacher's theological reflections, I argue that the covenant is actually a conclusion, a consequence of Schleiermacher's principles, commitments, and arguments.

If the eternal covenant is a result, it can explain all that other interpretations explain while also suffering from fewer problems. It can explain how authorities are negotiated non-arbitrarily, yet it needn't demarcate science from non-science, nor indeed make any *a priori* decisions on the disciplinary propriety of claims. In fact, it needn't explain the covenant as an agreed-upon political arrangement *at all*. Likewise, if the covenant is a result, it also needn't have its Reformation roots in the authority given to natural science *as a discipline*. Nevertheless, how the covenant has its basis in the Reformation can be explained on other grounds, namely, the doctrine of God.[27] And, if the covenant is a *result* of argumentation, it can enjoy all the same advantages of the accommodation model. It can explain how natural science can bear on religious commitments. It can explain why Schleiermacher rules out supernatural intervention as a mode of divine activity. It can explain why faith can make some claims over and above natural science. And most importantly, it can explain why the eternal covenant is properly called an "*eternal covenant*" at all, why it belongs in *The Christian Faith*, and why *The Christian Faith* does not merely announce the covenant but contains argument after argument in support of it. This last point is crucial; the eternal covenant is unintelligible apart from its basis in Schleiermacher's concrete claims and commitments. Many of these are explicitly theological and are not found in the *Letters*, but in *The Christian Faith*. And so, it is that work to which we now turn.

---

27 Schleiermacher must have been aware of the limits of such a claim—hence the disclaimer-like use of "basis." Nevertheless there is *substantial* basis for it. See, for instance, Susan Schreiner, *The Theater of His Glory: Nature and the Natural Order in the Thought of John Calvin* (Grand Rapids: Baker, 1991).

The Eternal Covenant in *The Christian Faith*

In discussions of the eternal covenant, *The Christian Faith* is often regarded as a showroom. It is seen as displaying Schleiermacher's proposal and its consequences. This suggests that *The Christian Faith* merely demonstrates—if it demonstrates anything at all—that there is at least one set of Christian beliefs that are compatible with the burgeoning success of natural science. This view is incomplete. Instead, Schleiermacher also supports the eternal covenant in *The Christian Faith* with a host of claims and commitments supported by argumentation.

These claims, commitments, and arguments are not only valid if one begins as Schleiermacher begins. One needn't accept Schleiermacher's starting point in consciousness to feel their force. Instead, the starting point for many of his claims is, for instance, simply God's aseity. And while Schleiermacher finds warrant for this commitment in the feeling of absolute dependence, one could easily find this same premise plausible for other reasons, as, indeed, many other theologians have.[28] The same holds true for other commitments. As we will see, Schleiermacher holds that no action is intelligible apart from an end for which to act. And, while in *The Christian Faith* he applies this principle to divine action, there is no reason to think that such a principle holds only when one is a Christian, let alone only after one has subscribed to Schleiermacher's starting point in consciousness. Since many of these commitments are not exclusive to *The Christian Faith*, their validity is not dependent on subscription to Schleiermacher's theology in every detail.

This is important for two reasons. One, it means that arguments which share these premises in common can be borrowed *from* sources which do not share Schleiermacher's theology of consciousness. And two, they can be valid *for* others who do not share the same. The following chapters will be devoted in part to substantiating the former claim—that Schleiermacher borrowed distinct but mutually supporting arguments from Spinoza and Leibniz, in particular, to support the covenant. Since Schleiermacher *borrowed* these arguments, they don't strictly depend on his theology of consciousness for their basis (though they are certainly compatible with it).

The point of pressing this former point is to give teeth to the latter. Too few thinkers have truly judged the bite of Schleiermacher's arguments, or weighed what it would really take to overcome them. In part, this is due to Schleiermach-

28 See Anselm of Canterbury, *Monologion*, Ch. 3; Thomas Aquinas, *Summa Theologiae*, Ia, q.3, a. 4, resp.

er's apparently packaging his arguments within his distinctive approach through consciousness. This pairing, however, is not essential. And that is exactly what the historical origin of these arguments demonstrates. If Schleiermacher's arguments could be borrowed, they can also be transplanted. Their force cannot be avoided by objections to tangential aspects of his project. Instead, the arguments which result in the covenant, and so the covenant itself, have an enduring relevance independent of Schleiermacher's prolegomena.

This is also important for the interpretation of the covenant in its own historical context. Argumentation implies intent to persuade. As above, this counts against readings of the covenant as a starting point. And yet, also as above, it is difficult to persuade those with whom you share nothing in common. So how did Schleiermacher attempt to persuade those who seemed to have little sympathy for his solution?

The answer cuts to the heart of the eternal covenant: Schleiermacher premised the arguments he gave in *The Christian Faith* on principles shared by both the Christian faith and natural science. His arguments are designed to expose inconsistency and incoherence. But his arguments are not *to* principles, but *from* them. In this sense Schleiermacher's mode is argumentation by description.[29] Descriptions pressed to their limits and turned back in upon themselves expose the false, the inconsistent, and the incomplete for what they are. That is how Schleiermacher can truthfully say he is describing the actual piety of Christians, not only without describing the specific beliefs of ordinary Christians, but even as he offers argument after argument against common beliefs which are both biblical and traditional. The dilemmas Schleiermacher poses are *ex hypothesi.* They take the form of "If...then..." claims. And these very hypotheses are principles and commitments shared by both him and his interlocutors, theological *and* scientific. And since these principles are shared, arguments from them can be made both within disciplines and between them. In Chapter 2 we will see how many of the natural scientists of his day shared not only explanatory principles, but theological ones as well, and how these were key components of their *scientific* programs.

In *The Christian Faith,* Schleiermacher uses these principles to argue for a specific account of divine action, both in relation to the divine nature and in relation to the world. It is an account which weds the divine nature to how the world is comprised and what sort of things happen in it. He argues both from

29 Schleiermacher certainly proceeds by describing the actual piety of Christians, but not by *simply* describing that piety. See Brian Gerrish "Nature and the Theater of Redemption," in *Continuing the Reformation* (Chicago: University of Chicago Press, 1993), 206.

the teleological and necessary nature of such action that the world proceeds in an ordered course, without intervention or supplementation, by the powers and forces inherent in the natural world. God, in short, works solely through secondary causes. That means that everything that happens in the world is, in principle, a natural event, explicable in terms of natural causes. And as Dole has rightly argued, the contents of the natural world include free finite causes: that is, human agents.[30] Free actions are, in principle, as explicable as any other event. As neither divine nor human action take place within the world as anything other than as natural events, the covenant requires theology to recognize the explicability of *everything* in the natural world by natural causes in principle.

Schleiermacher's two most important theological arguments for the covenant derive from two different applications of a version of the principle of sufficient reason. One argument deploys this principle under the aspect of divine wisdom, the other under the aspect of divine power. The first argument comes from Leibniz and is the subject of Chapter 3. The second comes from Spinoza and is taken up in Chapter 5. The two arguments together are, in Schleiermacher's hands, mutually supporting: power and wisdom are inseparable. An account of one without the other is, at best, an incomplete account. Each argument has its origins in separate arguments by Leibniz and Spinoza respectively, but, in Schleiermacher's hands, they are fused into a single account in spite of their *prima facie* incompatibility. Negatively, these arguments make a compound case against absolute miracles on the basis of divine perfection. Positively, they amount to something much more.

Schleiermacher's argument is not only against miracles, but for the unity of explanations. He proceeds from the principle that everything is explicable, that for everything there is a reason why it is so. Under this principle, a thing is either explained by itself, or by something else. The Christian faith is committed to the notion that, in principle, there is one and only one *Whence* which ultimately terminates the explanatory series.[31] Absolutely everything is explicable in terms of this one *Whence.*[32] Ultimate reality is therefore a unity, and so any explanation of the world will have to be a unified account. That is, in the final analysis, Schleiermacher is committed to the unity of knowledge.

But, recall: Schleiermacher does not divide the natural world from its finite agents, their judgments, purposes, values, and activities. Nor does he separate divine purpose from divine power. Instead, both are indispensable aspects of

30 Dole, *Schleiermacher on Religion and the Natural Order*, 35–69.
31 *GL* §4.4, 56.1; *The Christian Faith*, 16–17, 229.
32 *GL* §48; *The Christian Faith*, 184–89.

a complete explanation. Therefore, any account of the world as a whole, of reality both ultimate and penultimate, considered as a unity, will never be complete unless it includes final causes—both natural and divine. Whatever this final unified account looks like in detail, it will include facts and values, power and goodness.

By submitting human and divine action to the principle of sufficient reason and the unity of explanations, the covenant does not merely avoid embarrassment, but guarantees the incompleteness of any account which explains the world or God in terms of efficient causes alone. Likewise, to argue from the explicability of the natural world to the denial of theological premises is to defeat the very premises which guarantee the world's explicability. The covenant is as mutually supporting as it is mutually defeating. This explains the sense of immunity which advocates of the segregation model have picked up on. Theology and natural science rise and fall together. This is also why there can be claims over and above natural science on the accommodation model: natural science does not deliver first principles, but proceeds from them. However, theology is not immune from conceptual accountability either. Schleiermacher is not merely proposing "reverse-accommodation": that science defer to the deliverances of theology. No, the eternal covenant is an effort to secure mutual accountability, to seek consistency and completeness, to coordinate principles and content, to shun the *ad hoc*, the arbitrary, and to settle for nothing less than the ultimate unity of explanations.

Because there is, in principle, an explanation why everything is the way it is —from stars to humans, elements to minds—the world on Schleiermacher's account is determinate. Everything happens as it does, and in no other way, because everything else is the way it is and no other way. As we will see in Chapter 5, this holds even for God. The divine essence necessarily is such as it is because it is explicable in terms of nothing else, but sufficiently explicable in terms of itself alone. There can be no other reason for God to be, or act a certain way, than the divine essence itself. Since God is a wholly self-sufficient explanation, and since God is the sole sufficient explanation for everything else, everything whatsoever is determinate from the divine essence down. Were any one thing in the world to be differently determined, God could not be the same God. But since the divine essence is as it is and no other, everything else is as well. And yet, at the same time, this does not compromise freedom, human *or divine*, but perfects it. In fact, on Schleiermacher's account, divine determinacy is not only compatible with divine freedom, but is a necessary condition of it. Because the divine nature is determinate, and because determinate effects follow necessarily from determinate causes, the world and everything in it follows as inevitably, as it does freely, from the divine essence.

Moreover, this world has a purpose: namely the divine self-communication, motivated by divine love. This love is mediated by the perfect God-consciousness of Jesus of Nazareth through the Church by the power of the Spirit. Because God's redemptive activity is a feature of the world, and God, by nature, cannot do other than God, in fact, does, God is essentially and necessarily love, and, in consequence, necessarily redeems the world through Christ in this way. Nevertheless, Christ, the Church, the Spirit, and the potency for God-consciousness in humans are all natural phenomena. They arise only within and through the world. They are therefore only explicable, themselves, in terms of the nature system within which they occur. The redemption, accomplished by Christ, regards the divine love. The world and its order, by which and through which that redemption is accomplished, regards the divine wisdom. Only together as a unity of love and wisdom is God's creation, and the redemption it includes, adequately described as a supremely beautiful divine artwork. Finally, since God, by nature, can act in no other way as regards either ends or means, the world as a unity of plans and purposes, as a whole which is supremely beautiful, is the absolute revelation of the divine essence. This claim and its support are the subject of Chapter 6.

The world as revelation of the divine essence is the theological warrant for the eternal covenant. It can be compared to the covenants of scripture because it is predicated on the constancy of the divine essence and on the unity of the divine character and purpose. It can be called eternal because the divine essence upon which it is based neither comes into being nor passes away, and so the covenant, like that essence, cannot be other than it is. And it constitutes a relationship based on the pursuit of common goods since both theology and natural science are together in the business of the investigation of the divine essence: theology specifies divine ends, natural science specifies divine means. Together, and *only* together, can they offer a complete account of the ordered love which makes the world the self-revelation of God.

The Aims and Approach of This Work

Finally, a word on my aims and approach is in order. For practical purposes I have restricted myself to Schleiermacher's claims in *The Christian Faith*, though I make note of how my reading connects to other works of Schleiermacher's in reply to potential objections in Chapter 7. I am convinced that a coherent reading of Schleiermacher's views across his many works can be given, though it is not my aim to give such a reading here, nor does my case in this work rely on explaining Schleiermacher's thought *tout court*. My explicit intention is, rather, to elucidate what *The Christian Faith* says internally, on its own grounds, by stated claim and by necessary inference.

Further, I have restricted myself to relating the eternal covenant to natural science alone despite the fact that Schleiermacher also had other *Wissenschaften* in mind. This was done for two reasons. The first is because a central question of the eternal covenant is how to relate divine action and the natural world. In contemporary terms, the weight of the eternal covenant is on questions related to natural science. Second, according to my reading of the account of divine action and the natural world in *The Christian Faith*, whether we emphasize natural science, or any other science, the results will amount to the much same thing. Indeed, my account of the place of natural science in Schleiermacher's thought points us to history and human action, and vice versa. In other words, I take it that the eternal covenant between theology and science applies to all sciences as well as it applies to natural science. Nevetheless, natural science is the focus of this work.

In contrast to other accounts, I have not primarily focused on the political or social background to Schleiermacher's eternal covenant. Nor have I focused on the genesis and development of Schleiermacher's thought over the course of his life. Instead, mine is a conceptual analysis that emphasizes Schleiermacher's arguments in their rigor and detail as they appear in the final edition of *The Christian Faith*. I did this in part because I found no other fully adequate account of Schleiermacher's claims and arguments on the relevant topics in this work while, at the same time, there are many excellent works on the social, political, and ecclesial circumstances of the eternal covenant, as well as on the development of Schleiermacher's thought.[33] I, therefore, felt at ease relying on the good scholarship of others for such background.

I have also deliberately avoided emphasizing the introductory sections of *The Christian Faith*, and have especially de-emphasized the epistemic role (in either authorizing or limiting claims) of Schleiermacher's method. Rather, I have approached *The Christian Faith* first with an eye to its concrete claims, and then only very secondarily, with an eye to the epistemic warrant for these claims in the introductory sections. The reasoning behind this approach is this: since both appear in the same work, whatever particular claims Schleiermacher makes must, minimally, be compatible with his approach. And since, as we will see, Schleiermacher held that many of his claims are not only *possible* but *necessary*, whatever Schleiermacher has to say in the introduction must minimally authorize, or at least be intended to authorize, the necessity of at least

33 Again, see Dole, *Schleiermacher on Religion and the Natural Order.*

those particular claims of his that he took to follow necessarily. In other words, by beginning with the dogmatic content of *The Christian Faith* we discover that Schleiermacher took the religious self-consciousness to *entail* specifics. And, that means that the religious self-consciousness must be the sort of thing that is susceptible to yielding determinate content. Like a word, the meaning of Schleiermacher's introduction is its use. I am convinced that we understand the controversial claims of his introduction to *The Christian Faith* best by looking to the particulars he took those remarks to authorize, rather than by focusing most on his putative definitions.

Because Schleiermacher's case is cumulative, my argument is also cumulative. The following chapters will unpack his argument only one facet at a time. At no one step will Schleiermacher's argument be so clear and complete that the rest of his position can be understood without reference to its other parts and the whole they form. At any given point, many questions and objections will suggest themselves that either have already been answered in a different context, or that will have to wait to be answered until later on. I, therefore, beg the reader's patience, and ask that they not judge the merits of Schleiermacher's case as a whole until they know the whole. Only if each of the following chapters is taken as steps in one common argument, naturally inseparable from one another, can Schleiermacher's case be adequately understood.

Chapter 2 The Science of Schleiermacher's World, the World of Schleiermacher's Science: Natural Science in the early Nineteenth Century

> "Let me proceed to explain to you, Socrates, the order in which we have arranged our entertainment. Our intention is that Timaeus, who has made the nature of the universe his special study, should speak first, beginning with the generation of the world and going down to the creation of man."
>
> - Plato, *Timaeus*

When Karl Barth, commenting on Schleiermacher's *Letters to Lücke*, concluded that, "*mutatis mutandis* the coming of Christ is similar to the formation of a new nebula,"[34] the bite of such a comparison would not have missed his audience. New nebulae are being formed all the time. Their formation is apparently unremarkable. If, as Barth sees it, the parallel Schleiermacher intends to draw is between the origin of nebula and the coming of Christ, then the *ordinariness* of both events looms large. If so, it is not hard to sympathize with Barth's dissatisfaction. Christ is as unremarkable as rotating gas and particles collecting in a vacuum.

Nevertheless, for all that, Barth gets Schleiermacher wrong on this point. But his mistake is not that he draws a parallel between the development of Christ out of the nature system and the natural development of nebulae. Schleiermacher means to make the comparison—though not between the banality of events. No, Barth's mistake, oddly, stems from misunderstanding *nebulae* in their early nineteenth century context. In consequence, the comparison is inexact and Barth misses Schleiermacher's point.

This might seem strange. What could be so important about nebulae in the early nineteenth century that could specify Schleiermacher's meaning so differently? Unfortunately, there is no ready answer to this question. There is, in general, a conspicuous silence on the part of Schleiermacher scholars on the science of Schleiermacher's day. Given Schleiermacher's stated intent to develop his the-

34 Karl Barth, *The Theology of Schleiermacher*, trans. Geoffrey Bromily (Grand Rapids, MI: Eerdmans, [1923/24, 1978] 1982), 205.

DOI 10.1515/9783110542301-002

ology in accord with the natural sciences in general and with the scientific doctrines of his day in particular, it is problematic that in nearly two hundred years no one has looked to the particular science mentioned and employed in *The Christian Faith* and the *Letters to Lücke*, or even the science of his day writ large, in connection with these works. This neglect has led to anachronism, and it was this anachronism which led Barth astray. If contemporary interpreters wish to avoid the same or similar mistakes, they must attend to the meaning and context of Schleiermacher's specific scientific commitments in their nineteenth century context.

Scientific commitments, however, might seem to be in short supply in the *Letters to Lücke* and *The Christian Faith.* Not so. For, although mentioned often in passing, and rarely for their own sake, a number of scientific doctrines make an appearance between the two works. Ranging from topics as diverse as the duration and extent of the cosmos and the continuing development of the world, to the evolution of new species and the extinction of old ones, Schleiermacher holds a host of specific scientific teachings in his theological work.

These teachings all have their sources in the science of the late eighteenth and early nineteenth centuries. Unfortunately, none of those sources are explicitly named. This, some might object, poses an insurmountable obstacle to any attempt to connect Schleiermacher's specific views with specific sources. But a quick glance at *The Christian Faith* will reveal a dearth of citation in general—even among known influences. Spinoza, for example, appears nowhere in the index and is not named in the text as a source. And yet, few Schleiermacher scholars have any doubts that many of Spinoza's ideas made their way into *The Christian Faith.* In this case, scholars don't rely on direct citation to warrant their claims. Spinoza's ideas are, instead, plainly recognizable to the familiar reader. They could be picked out of a lineup. In many cases, they could come from no one else.

The same recognizability holds for Schleiermacher's scientific ideas. Minimally, their origin can be narrowed down to one of two or three likely candidates, or some combination of candidates. In some cases the uniqueness and specificity of Schleiermacher's views appears to suggest only a single best option. But in nearly every case the scientific doctrines he holds are specific enough, and their sources prominent enough, that the influence of major figures is as recognizable as that of Kant, or Spinoza, or Plato. The reason Schleiermacher scholars have not recognized these ideas as clearly as some of Schleiermacher's philosophical or theological sources is simply due to the fact that many are, like Barth, unfamiliar with the natural science of the early nineteenth century.

Knowledge of the exact content of these scientific teachings is, however, not necessarily to be gained for its own sake. It might be significant in and of itself, but it might not be. What matters most is that much of the science was decided by, or at least made more plausible in light of, general principles of operation of the universe. If the world works in certain general ways, certain specific operations are more or less believable. On the other hand, particular discoveries also count toward the plausibility of a certain view of the world as whole. And, in fact, the particular teachings which follow imply just that.

The aims of this chapter are threefold: one, to identify, wherever possible, exactly what specific natural scientific beliefs Schleiermacher subscribed to; two, to show the formal similarities in how these beliefs were supported through an examination of the arguments offered by the natural scientists under question; and three, to make plain that natural science and theological commitments were already organically related and mutually accountable *in the natural science of the day.* Thus, when Schleiermacher argues for the eternal covenant from theological first principles he is not arguing against natural science, but is rather joining a conversation that is already well underway. Moreover, since, as we will see, theology and science are not demarcated even in the natural sciences, we should not expect that they be cordoned off from one another in Schleiermacher's theology. Instead, Schleiermacher is joining a long tradition, shared by much of the natural science of his day, of coordinating higher and lower principles, and coordinating principles with content in a holistic account of reality.

Schleiermacher and His Contemporaries on the Duration, Extent, and Evolution of the Cosmos

In his *Letters to Lücke*, Schleiermacher warned that,

> [W]e must learn to do without what many are still accustomed to regard as inseparably bound to the essence of Christianity. I am not referring to the six-day creation, but to the concept of creation itself, as it is usually understood [...] How long will the concept of creation hold out against the power of a world view constructed from undeniable scientific conclusions that no one can avoid[...].[35]

If Schleiermacher's warning was clear, the threat and its solution were not. He gives no more detail in the *Letters*, but instead turns to the subject of miracles.

---

35 Schleiermacher, *Letters to Lücke*, 60 – 61.

Yet, though Schleiermacher's reference is vague, when combined with his treatment in *The Christian Faith*, the subject comes into focus.

First, it is clear what giving up the concept of creation cannot entail: giving up the absolute dependence of the world on God. That dependence is the indispensable feature of any God-world relationship and is essential to the Christian consciousness of God. So whatever giving up creation "as it is usually understood" means, it cannot mean giving up either absolute dependence or the creator-creature distinction which that dependence implies. But if creation in its broadest sense (as elaborated in *GL* §§36–41 and §46) cannot deny dependence, what can it exclude? There is only one possibility: namely, that giving up the usual understanding of creation might mean giving up a beginning.[36] Schleiermacher holds that, "In general the question of the origin of all finite being is raised not in the interest of piety but in that of curiosity, hence it can only be answered by such means as curiosity offers."[37] That is, the question of a beginning to the world is a question reserved for scientific inquiry. By implication, it can be answered in the negative as easily as in the affirmative. And if the world has no beginning, it has an infinite past. Thus the world conceivably has an infinite duration. And yet:

> [I]t still does not by any means follow that the temporal existence of the world must reach back into infinity, so that no beginning can be thought of [...] On the other hand, we need not be anxious lest, if the world is given no beginning or end, the difference between divine causality and causality within the natural order should be cancelled, and the world be as eternal as God. On the contrary, the eternity of God remains none the less unique, since the antithesis between the temporal and eternal is not in the least diminished by the infinite duration of time.[38]

Although no anxiety is warranted, its *potential* source is clearly the looming possibility of a world of infinite duration.

Back in the *Letters to Lücke*, Schleiermacher mentions another infinite: the infinite extent of space. This claim is included in Schleiermacher's nebular comparison. As he says in passing, "If science must admit the possibility that even now matter is beginning to form and rotate *in infinite space* [...],"[39] and so on. We will return to the larger context of the quotation below, but, for now, this one phrase deserves close examination. The notion of infinite space is assumed. It is almost a side comment and neither what proceeds, nor what follows, depends

---

36 *GL* §39; *The Christian Faith*, 148–49.
37 *GL* § 39.1; *The Christian Faith*, 149.
38 *GL* § 52.1; *The Christian Faith*, 204–205.
39 Schleiermacher, *Letters to Lücke*, 64, Emphasis added.

on it. Schleiermacher seems simply to take for granted that the best science of the day holds to the infinite extent of the cosmos. In and of itself, belief in the infinite extent of the world is far less remarkable than its infinite duration but is, nevertheless, noteworthy—and especially so in *combination* with the infinite duration of the world.

If the infinite duration (and, possibly, extent) of the world bears on that "usually understood" sense of creation which, Schleiermacher warns, Christians might well have to do without, why think this threat is pressing? Since the sheer possibility of the infinite duration of the world has long been admitted in Christian theology,[40] the source of Schleiermacher's urgency is unlikely a new philosophical or theological argument. Powerful philosophical arguments for the infinite duration and extent of the world were, by Schleiermacher's day, made as recently as Spinoza, and yet those arguments caused little anxiety among the theologians—at least little enough that theologians did not feel obliged to change their views in light of them. A new philosophical argument *alone* is an insufficient explanation. Rather, it is almost certainly an issue pressed by natural science. But, if the infinite duration of the world is suggested by natural science, what science does Schleiermacher have in mind?

Here again, the novelty and urgency of the question are telling. If the science was not at least relatively new, the issue could not be so novel. We would have expected earlier theologians to deal with it in more detail. Instead, as Schleiermacher points out, the theologians of the Reformation and their successors seem to offer rather simplistic matter-of-fact accounts of creation from the Genesis narrative.[41] Even relative to his immediate forerunners, Schleiermacher is dealing with something new. So the science involved must be at least relatively recent, roughly contemporary with Schleiermacher. And so, something both new and pressing is required to account for Schleiermacher's warning to prepare to give up "the concept of creation itself, as it is usually understood," something likely derived largely from the natural science of the time, or at least something supported by it.

If Schleiermacher merely held to the infinite duration and extent of the cosmos, that alone would be insufficient to narrow down his likely influences. Fortunately, Schleiermacher claims much more. When he wrote in the *Letters to Lücke* that, "If science must admit the possibility that even now matter is beginning to form and to rotate in infinite space,"[42] etc., he implies subscription to a

---

40 See Thomas Aquinas, *On the Eternity of the World*, trans. Cyril Vollert (Milwaukee: Marquette University Press, [1964] 2003).

41 *GL* §40.2; *The Christian Faith*, 150–51.

42 Schleiermacher, *Letters to Lücke*, 64.

series of scientific doctrines which individually, and especially collectively, specify his cosmology. Barth was, indeed, correct when he associated this view with the formation of nebulae. Though Schleiermacher does not use the word, we can plainly recognize the notion. But care must also be taken not to rush past the details of Schleiermacher's particular view, details that carry with them a good deal of content.

First it is important that something is *being formed*. That is, Schleiermacher is specifying a process of becoming. Since Schleiermacher argues that if science must admit this, "then it must also admit that in the realm of spiritual life there is an appearance that we can only explain as a new creation, as the beginning of a higher development of spiritual life," the thing in process of becoming must also be some kind of new creation, and, if the parallel holds, some kind of higher development. But this novelty, and this notion of development itself, are themselves tied to the phrase "even now." That is, this *development* is taking place *even now*. The development with which Schleiermacher is concerned is an ongoing developmental process, bringing a new sort of thing into existence.

Second, what is being formed in the cosmos is *rotating*. There is, of course, no parallel to a "higher development of spiritual life" in the idea of rotation. Instead the import of such a phrase is merely cosmological. It signifies either an instrument of formation or the form which that becoming takes. In either case, this phrase is helpful in locating Schleiermacher's view of formation of new matter in space as falling broadly within the *vortex* tradition.[43] That is, matter aggregates to other matter like water drains from a tub, by being drawn to a central point by some force. This matter does not rush immediately to the center of attraction, but again, like draining water, rotates and approaches the center only indirectly. As a view of the formation of celestial systems and objects, this doctrine is rather unremarkable. The general idea had been proposed by Descartes some two hundred years prior.[44] Instead, formation by vortex is noteworthy in specifying Schleiermacher's view on this matter beginning to rotate and form as a reference to nebulae, and by specifying the notion of nebulae he has in mind.

The modern notion of nebulae should not be taken for granted in Schleiermacher's science. That he, in fact, held to a modern notion of nebulae is a remarkable feature of his account. But in the late eighteenth and early nineteenth century *what* a nebula was, and what it signified, were hotly debated issues. The

43 I say "broadly" because I do not mean that Schleiermacher or the cosmologists of his day necessarily subscribed to the details of that tradition, for instance, that the æther was the mechanical medium for the formation of vortices rather than gravity.
44 See René Descartes, *Le Monde* (1667).

word simply derived from the "cloudiness" of certain stars, that is, from the apparent inability of the best telescopes of the day to crisply resolve their image. A great deal of controversy surrounded what this cloudiness implied. But, by Schleiermacher's day, the astronomer William Herschel (1738–1822) had found, using newer, more powerful, telescopes, that at least some nebulae were resolvable into clusters of distinct stars, "plainly to be distinguished with my large instruments, or such nebulous appearances as might be reasonably supposed to be occasioned by a multitude of stars at a vast distance."[45] And yet, importantly, although some nebulae came into focus, others remained unresolvable.

Herschel's observations suggested a great deal beyond resolving a minor controversy. If nebulae were resolvable into their component stars that meant that those nebulae had to be at an enormous distance from Earth in order for their light to diffuse enough to cause such a cloudy appearance. In other words, whatever these nebulous clusters are, they are faint. But if, as appears, they are composed of *stars*, then in order for these extremely bright objects to be so faint, they must be correspondingly distant. Moreover, in order for a *cluster* of such stars to appear as a single body, individual stars must be a small speck in a faint object. In short, the sheer immensity of space was more and more apparent. And it was this growing sense of immensity which provided an easy stepping stone to infinity.

But that was not the most important implication of Herschel's discovery. His second, and most important insight was that nebulae *developed*. Important in itself, and a theme to which we will return below, that development also implied a nearly unimaginable *duration* of the universe.

Though Herschel was able to resolve some nebulae into their component parts, he was not able to resolve all in the same way. "Many of the nebulae," said Herschel, "had no other appearance than whitish cloudiness, on the blue ground upon which they seemed to be projected [...]."[46] And yet, these were not so much a distinct class of nebulous object, but especially-cloudy nebulae on a continuum of haziness to distinctness. This spectrum was filled out by Herschel's further observations of nebulae somewhere in between, consisting of a distinct bright center and yet still wreathed with a glowing cloudiness. He concluded that there were only two explanations for these phenomena: one, that the cloudiness consisted of normal-sized stars and the bright "star" at the center was

---

45 William Herschel, "On Nebulous Stars, Properly So-Called," in *Philosophical Transactions of the Royal Society of London*, 1791, 71.
46 Herschel, "On Nebulous Stars," 74.

something so enormous that it "must far exceed the standard of what we call a star,"[47] that it must be something "of a nature totally unknown to us"[48]; or, conversely, that the "star" at the center of such nebulae was a star in the ordinary sense, but that whatever it was that composed the 'nebulosity' must be "very small and compressed,"[49] and therefore no star either. The stars at the center of nebulae and whatever made up their glowing "*chevelure*" could not both be "stars" in the same sense and of the same scale.

From our historical vantage, it appears that both options were, in important ways, right. Herschel probably observed some individual nebulous stars as well as some galaxies, and so quite possibly did identify some objects as stars surrounded by "nebulous fluid" which were actually galactic cores surrounded by ordinary stars "very small and compressed" by comparison. But, Herschel himself had a decided preference for the latter option—that the nebulosity of nebulae was due to "a shining fluid"[50]—and he was not incorrect in this identification of normally sized stars as, at least some of the time, their core constituent. He argued that, in these cases, it is "highly probable" the nebulous star and the nebulosity surrounding it "are of the same nature."[51] From this insight, it was not far to his revolutionary hypothesis. If star-like, then, Herschel argued, the glowing nebulous material must be self-luminous. And "If, therefore, this matter is self-luminous, it seems more fit to produce a star by its condensation than to depend on the star for its existence."[52] Nebulae, not merely the effects of stars, were instead taken to be their causes.

What Herschel implied, Pierre-Simon Laplace (1749 – 1827) made explicit. "Herschel," Laplace said, "while observing the nebulae by means of his powerful telescopes, traced the progress of their condensation, not on one only, as their progress does not become sensible until after the lapse of ages, but on the whole of them, as in a vast forest we trace the growth of trees, in the individuals of different ages which it contains."[53] Just as the growth of a tree is too slow to observe from start to finish, so the growth of stars is far too slow to see in an individual star. Instead, each star, like each tree in a forest, is, by analogy, a snapshot of their common development. Of course, this supposes

---

47 Herschel, "On Nebulous Stars," 83.
48 Herschel, "On Nebulous Stars," 83.
49 Herschel, "On Nebulous Stars," 83.
50 Herschel, "On Nebulous Stars," 83.
51 Herschel, "On Nebulous Stars," 85.
52 Herschel, "On Nebulous Stars," 85.
53 Pierre-Simon Laplace, *The System of the World*, trans. Henry Hickman Harte (Dublin: [1796] 1830), 336.

that stars are, like trees, of a kind. But Herschel, and Laplace after him, considered this matter settled.[54] Once stars are known to be of a kind, and the *chevelure* of nebulae are also thought to be of the same nature as the star they surround, it is simple to infer that all stars find their origin in nebulae. The various stages of "condensation" from nebulous fluid into full stars became windows into the common lives of all stars.

Laplace did not stop there, however. He turned his historical analogy into a predictive, or, more exactly, a perpetual one. When we see a whole forest of young and old trees in an instant, and infer that mature trees grew from saplings, we not only make an inference about the origins of mature trees, but also the future of saplings. Laplace extended this same inference to nebulae. "The nebulae," he said, "classed in a philosophical manner, indicate, with a great degree of probability, their *future* transformation into stars, and the anterior state of nebulosity of existing stars."[55] That is, new stars are being formed all the time from existing nebulae. If one subscribed to universal gravitation, as all these astronomers did, the accretion of matter, nebulous or otherwise, was uninterrupted. And if one subscribed to the *practical*, if not actual, infinite duration of the universe, as these astronomers also did,[56] the accretion of nebula must have been going on for an immense period of time. With no reason to suspect an exception, and an infinite or practically infinite amount of past time during which the formation of nebulae had not yet ceased, it seemed most plausible that the universe was *presently* and *continually* forming new nebulae which themselves formed new stars.

In the end, this resulted in the most important feature of the nebular hypothesis in relation to Schleiermacher's comments: namely, that the cosmos is an *evolutionary* cosmos, a cosmos in which *even now* things are coming to be which had only before existed in nascent potential. Herschel's discoveries, combined with Laplace's synthesis, made the *celestial-evolutionary nebular hypothesis*[57] of the formation of the universe a live option. When Schleiermacher said that "even now matter is beginning to form and to rotate in infinite space," and that this is the means of a general process of becoming in the universe,

---

54 William Herschel, "On the Nature and Construction of the Sun and Fixed Stars," in *Philosophical Transactions of the Royal Society of London*, 1794, 68.

55 Laplace, *The System of the World*, 336, Emphasis added.

56 Herschel "On Nebulous Stars," 87; Laplace, *The System of the World*, 335–336.

57 The *celestial*-evolutionary variant did not necessarily imply biological evolution or vice versa. Of course, the two were amenable to one another, and perhaps mutually suggestive, but still not strictly necessary.

he was deploying the cosmology of his contemporaries in support of this broader view of universal development.

The celestial-evolutionary nebular hypothesis found support in the work of Herschel and Laplace, but it did not originate with them. In fact, that hypothesis in its most complete version was not proposed by a scientist or mathematician. Instead, almost fifty years before Herschel, it was developed by Immanuel Kant.[58]

Kant elaborated his version of the evolutionary nebular hypothesis in 1755 in his *Allgemeine Naturgeschichte und Theorie des Himmels* (hereafter given in translation as *Universal Natural History and Theory of the Heavens*[59]). In this work, Kant proposed a grandiose cosmogony which explained, using aspects of Newtonian mechanics, a system of the world's generation—and continuous development—in which analogous nebular formations were the basic *explanans*. In other words, Kant proposed an evolutionary nebular theory of the formation of the *entire* universe, not just of single stars.

The status of this work as natural science is controversial. In his extensive introduction, Stanley Jaki calls the *Universal Natural History* more a teleological-philosophical essay than a work of natural science. [60] According to Jaki, Kant was patently novice in Newtonian mechanics, and even committed simple mathematical errors.[61] Moreover, the *Universal Natural History* was passed over in silence in important German sources even as Herschel's discoveries were made public.[62] Jaki's conclusion is that Kant's *Universal Natural History* has been given far more credit than it deserves, owing largely to the nineteenth century neo-Kantian revival.[63] Instead of a work of great daring and originality it is instead, according to Jaki, the mere fantasy of an arm-chair cosmologist.

Obstacles, therefore, remain in connecting it to the science of Schleiermacher's day—in spite of the *Universal Natural History* being presented, *contra* Jaki, as an important and novel theory in textbooks published well after Jaki's introduc-

---

58 For a less ambitious immediate precursor to, and inspiration for, Kant's theory see Thomas Wright, *An Original Theory or New Hypothesis of the Universe* (1750). For an immediate related successor see J. H. Lambert, *Cosmologische Briefe über die Einrichtung des Weltbaues* (1761).

59 Immanuel Kant, *Universal Natural History and Theory of the Heavens*, trans. Stanley Jaki (Edinburgh: Scottish Academic Press, 1981).

60 Kant, *Universal Natural History*, 10.

61 Kant, *Universal Natural History*, 12.

62 Kant, *Universal Natural History*, 46–47.

63 Essentially, Jaki's whole introduction amounts to a running criticism of Hastie's translation and promotion of the *Universal Natural History*. See especially Kant, *Universal Natural History*, 51–71.

tion.[64] The most pressing question is Schleiermacher's access to Kant's ideas, not necessarily access to the text of the *Universal Natural History.* If Schleiermacher couldn't have known them, he couldn't have had them in mind, irrespective of their quality.

However, Kant's ideas in the *Universal Natural History*, though perhaps silently passed over in scientific circles, were nevertheless alive elsewhere—even according to Jaki himself. Specifically, two essays published in 1792 and 1793 by Johann Christian Schwab attacked aspects of Kant's theory.[65] From here the distance to Schleiermacher is short. Schwab was a friend and collaborator of J. A. Eberhard's, in whose anti-Kantian journal he published his two essays.[66] Eberhard, who will make an appearance again in Chapter 3, was one of Schleiermacher's most admired and influential teachers at Halle. There is no question then that Schleiermacher had access to Kant's ideas in the *Universal Natural History.* Anyone keeping abreast of the astronomical developments of the day would have seen in Herschel and Laplace's work growing confirmation of the general outline of Kant's nebular hypothesis. Of course, the best argument for the influence of Kant's ideas in this regard is the connection of his claims to what Schleiermacher entertained.

Kant held, more strongly than either Herschel or Laplace, that both the duration and extent of the cosmos were infinite. Neither Herschel nor Laplace were comfortable with claiming the universe is *actually* infinite in either respect. Instead, Herschel is representative of both when he says that, "the number of the emitting bodies is *almost* infinitely great, and the time of the continual emission *indefinitely* long [...]."[67] Kant, on the other hand, entitles the seventh section of the second part of his book, "Of creation in the entire extent of its infinity, both according to space and time."[68] And while he is, perhaps, less than consistent in its consequences,[69] he is committed to these views by his own premises.

---

64 See, for instance, Frederick Gregory, *Natural Science in Western History* (Boston: Houghton Mifflin, 2008), 188.

65 Kant, *Universal Natural History*, 45.

66 Kant, *Universal Natural History*, 45 n. 45; See Schwab, "Prüfung der Kantischen Hypothese von dem mechanischen Ursprung des Planetensystems," *Philosophisches Archiv*, 1, 2 (Berlin: 1792), 1–36; "Prüfung der Kantischen Hypothese Ursprung des Ringes des Saturn, und die Berechnung der Achsendrehung dieses Planeten," *Philosophisches Archiv* 1, 2 (Berlin: 1793), 1–21.

67 Herschel "On Nebulous Stars," 87, Emphasis added.

68 Kant, *Universal Natural History*, 95.

69 Kant sometimes speaks, for instance, of a 'beginning' even while insisting the universe has an infinite past. Elsewhere Kant seems instead to suggest the universe has a finite past, but an infinite future, thus potentially resolving this contradiction. See *Universal Natural History*, 151–152.

Kant's commitment to the infinite duration and extent of the cosmos follows far more from the principle of plenitude than from astronomical observation:

> The doctrine, which we have proposed, opens for us an outlook into the infinite field of creation and offers a picture of God's work which is appropriate to the infinit[y] of the Great Artificer [...] There is no end here but an abyss of true immeasurability in which all ability of human concepts sinks even when it is elevated by the help of the science of numbers. The Wisdom, the Goodness, the Power, which reveals itself, is infinite and is fruitful and operative in the same measure; the plan of its revelation must therefore, just as it is, be infinite and without limits.[70]

The universe must be infinite in proportion to the power of its Creator.

Kant also, more imaginatively than Herschel or Laplace, held that the universe continues to evolve, even now. This is in part due to his insistence on the infinite duration and extent of the universe, and in part due to other factors which we will discuss below. For Kant, this kind of nebular formation takes place not only for nebulous stars, nor also only for planets revolving around suns, but even for *systems of stars*, i.e. galaxies, and finally for the universe as a whole. In fact, as he proposes, the nebulous stars are themselves other Milky Ways.[71] They are far too large to be single objects, but,

> Far more natural and comprehensible is [the view] that they are not single great stars, but systems of them whose distance present them in so narrow a space that the light, which is unnoticeable from each individually, issues in their immeasurable multitude into one uniform faint glimmer.[72]

These immense systems of stars are continually being formed across infinite space. "Creation is not the work of a moment," Kant says:

> After it has made the start of bringing forth of an infinity [of] substances and matter, it is efficient throughout an entire sequence of eternity with an ever growing degree of fruitfulness. Millions and entire mountains of millions of centuries will flow by, within which always new worlds and world-orders form themselves one after another [...].[73]

---

70 Kant, *Universal Natural History*, 108.
71 Kant, *Universal Natural History*, 106–107.
72 Kant, *Universal Natural History*, 107.
73 Kant, *Universal Natural History*, 154.

Again, Kant reiterates, "Creation is never completed. Though it once has started, but will never cease. It is always bringing forth more scenes of nature, new things and new worlds."[74]

Kant's *Universal Natural History* contains both views which are most characteristic of Schleiermacher's cosmology in the *Letters to Lücke* and *The Christian Faith:* an *evolving* universe, hypothetically, of *infinite* duration and extent. The combination of these views cannot be found in either Herschel or Laplace, who restricted themselves to more modest claims. Of course, Kant's grandiose hypothesis would have been lent renewed plausibility by anyone reading Herschel and Laplace. They provided additional reasons, especially observational support, to think something like Kant's view was right. The combination of Kant's imaginative cosmogony, its actual infinites and wholly evolving universe, *together* with Herschel and Laplace's more empirically supported conclusions a half-century later, likely account for why Schleiermacher found the possibility of giving up a beginning so pressing and why he found it so plausible that "even now matter is beginning to form and rotate in infinite space."

Eighteenth Century Cosmology and Its Commitments

In calling Kant's essay a "reformulation of teleology on the basis of mechanistic science, an obviously philosophical enterprise,"[75] Jaki means to distinguish Kant's efforts from natural science proper which is, by implication, more empirically grounded. Unfortunately for Jaki, Kant may be wrong, but there is nothing different about his efforts in kind. And even more unfortunately, not only Kant but also Herschel, Laplace, and anyone engaged in the cosmological debates of the time, made heavy use of non-empirical beliefs, many of which were explicitly theological in nature. Not only are these beliefs crucial for understanding the arguments these thinkers made, but they will turn out to be some of the very same commitments for which *and from which* Schleiermacher argues in subsequent chapters.

Kant, Herschel, and Laplace, all make heaviest use of argument from analogy. Examples are too plentiful to exhaust, but include the following. The first is Kant's analogical support derived from the rings of Saturn and the plane of the ecliptic of our solar system for hypothesizing about the flattened disk shape of the Milky Way, and from there to other celestial systems. As Kant says, reflecting

---

74 Kant, *Universal Natural History*, 155.
75 Kant, *Universal Natural History*, 10.

on his proposal, "if conjectures, in which analogy and observations agree perfectly to support one another, have the same dignity as formal proofs, one must hold the certainty of these systems to be demonstrated."[76] Herschel's argument too, depends on analogy. "A well connected series of objects, such as we have mentioned above, has led us to infer," he says, "that all nebulae consist of stars. This being admitted, we were authorized to extend our analogical way of reasoning a little farther."[77] In the third place, as we saw above, Laplace's reasoning from trees in a forest depends on a *double* analogy: first, that by which we analogically infer that young trees resemble the youth of old trees and old trees the future of young ones; and second, that by which forests of trees resemble a field of stars and nebulae. And finally, it is only through the service of analogical reasoning that Schleiermacher compares the coming of Christ to a nebula. Analogical reasoning is put to ubiquitous use by all parties.

Analogy itself involves the conceptual synthesis of continuity and difference applied to diverse things and to the universe as a whole. Two things can only be two *different* things if there is actually some difference, and yet they can only be compared if there is some continuity underlying that difference. The synthesis of continuity and change is *development*, the red thread which connects all these cosmological theories, and, which we will see below, links them to life. Even as these thinkers became convinced, as Laplace put it, "that nature is far from being always every where the same,"[78] they became equally convinced in the explicability of this diversity through orderly operations inherent in nature and explicable solely through the natural order. Interestingly, this growing diversity actually had the effect of strengthening claims for continuity. The success of particular analogical comparisons, relying implicitly on a synthesis of continuity and difference, underwrote the growing explanatory power of *development* at large.

Kant, Herschel, and Laplace had a decided preference for explaining this development through the concept of *mechanism*. However, as we will see, their specific use of the term "mechanism" might mislead. It was not primarily used in contrast with *organism*, but rather with divine intervention, and with occasionalism—two subjects that we will explore in detail in subsequent chapters. Put positively, "a mechanical doctrine,"[79] in this sense, is, according to Kant, the belief "that the world recognizes for the origin of its constitution a mechanical de-

76 Kant, *Universal Natural History*, 107.
77 Herschel "On Nebulous Stars," 74.
78 Laplace, *The System of the World*, 334–335.
79 Kant, *Universal Natural History*, 168.

velopment unfolding from the general laws of nature."[80] To subscribe to "the mechanical doctrine" in this context was to take a side in an ongoing dispute. As Kant explains:

> It still remains to be decided whether the plan of the constitution of the universe has been deposited by the highest intellect into the essential dispositions of eternal natures and planted in the general laws of motion so that it may evolve from them in a manner fitting a most perfect order; or whether the general properties of the constituent parts of the world have the complete inability to harmonize and not [even] the slightest trend for union, and need therefore a foreign hand [so as] to move into that compactness and coordination which permits perfection and beauty to be seen [everywhere in nature].[81]

That is, as Kant presents it, mechanism has become one of two very general options in thinking about the development of the natural world and God's role in it. "The mechanical doctrine" holds that God has implanted in both the natures of things, and in the laws of nature as whole, the natural disposition to evolve into the order in which we behold them. In contrast, the other, unnamed, view denies just this. Instead, the alternative to mechanism in this context is the view that nature must be ordered extrinsically by the hand of God.

In arguing for mechanism, Kant is taking a clear stance on both the constitution of the world and on divine action. Both continuity and difference in the natural order are to be explained solely in terms of powers and forces inherent in the world. Whatever happens in the world is always an actualization of an inherent natural potency of the universe at large. Of course, in his *Universal Natural History* Kant is keenly aware of the obvious theological objections to this notion of mechanism. In fact, his cosmology is so thoroughly theological that, better than Jaki's description of it as a teleological-philosophical essay, it more accurately stands in the aptly named "Astro-theology" tradition.[82] And yet, at the same time, it is no less scientific for it. Kant genuinely set out to get the formation of the universe right, and he argued for his conclusions from a host of commitments, some of which happened to be theological. But, as it was alternative *theological* beliefs which principally stood in his way, to propose mechanism as an explanatory strategy *required* one to oppose other conceptions of divine action which contradicted it. For those who believed in God (which was true of most, and expected of all), mechanism was, first and foremost, a way of conceiving of divine action, and only subsequently a strategy for explaining the op-

80 Kant, *Universal Natural History*, 170.
81 Kant, *Universal Natural History*, 169.
82 See William Derham, *Astro-theology: or, A Demonstration of the Being and Attributes of God from a Survey of the Heavens* (London: 1715).

erations of the universe. In order to support mechanism, he had both to defend it against reduction to atheism and to press the burdens which attended his opponents views back upon them.

The theological burden to mechanical explanation was so great that Kant begins the *Universal Natural History* by addressing it before setting out *any* of his hypotheses about the world or advertising their strengths. Speaking as his opponents, Kant objects:

> If the world-edifice with all [its] order and beauty is only the effect of a matter abandoned to the universal laws of motion, if the blind mechanism of the forces of nature knows how to develop itself so splendidly from the chaos and to reach by itself such a perfection, then the demonstration of [the existence] of a divine Author, which one derives from the consideration of the beauty of the world-edifice, is wholly invalidated, nature is sufficient to herself, the divine government is unnecessary, Epicurus lives again in the midst of Christendom, and an unholy philosophy tramples underfoot the faith which provides for it a clear light to enlighten philosophy itself.[83]

The theological burdens of his cosmology are not restricted to his particular explanation, but rather to his broader explanatory program. And Kant is not without sympathy for his hypothetical objector. Quoting him again at length:

> The defender of religion is worried that these harmonies, which can be explained from a natural disposition of matter, should prove the independence of nature from divine providence. He admits very plainly that if one can discover for all order in the world-edifice natural grounds, which can produce all that order from the most universal and essential properties of matter, then it becomes unnecessary to appeal to a highest government.[84]

If all that happens in nature can be explained naturally, some fear that nature as a whole is then sufficient to explain itself. And, if the natural order is wholly sufficient to explain itself, little work remains for God.

However, Kant, remained committed to the mechanical development of the cosmos in the face of such objections. In the place of supplementation, he argued from both divine wisdom and divine power that such objections bear their own burdens. As Kant says:

> But I answer: if the universal laws of matter are similarly a consequence from the highest plan, then presumably they can have no other destinations than the ones which themselves aim at fulfilling the plan which the highest Wisdom has set to Itself; or if this is not so, should not one be lured into the temptation of believing that at least matter and its univer-

83 Kant, *Universal Natural History*, 81–82.
84 Kant, *Universal Natural History*, 82–83.

> sal laws were independent and that the wisest Power, which knew how to make use of them so splendidly, is though great, yet not infinite, though powerful, yet not all-sufficient?[85]

These arguments for the completeness and sufficiency of nature will remain a theme in the following chapters. They were central to what Kant called the 'mechanical' operation of the universe and what Schleiermacher will describe as its organic development. For now, what is especially important to note is this: that one and the same argument for continuity in the development of the universe was appealed to *in the scientific literature* to which Schleiermacher's views most closely correlate. Thus, when Schleiermacher appeals to the science of his time, he is already appealing to the results of investigations which seek to explain the universe under the auspices of the *developmental hypothesis*, an explanatory principle, which proposed to explain everything in the natural world as having developed entirely through natural processes. And it is the manifest success of this principle and others in explaining the world as it appears that lends scientific credence *to these principles*, and not merely to the particular explanations which use them, even as Schleiermacher will go on to support these mediating principles with higher ones in the doctrine of God.

Schleiermacher on the Evolution of Life

For Schleiermacher the developmental nature of the universe is not limited to stars and planets but also includes the evolution of life. Not only does this follow the best science of the day, but, as we will see, it is also a natural extension of the principle of development which his cosmology implies. And further, as we will see, it is to this principle which the natural philosophers studying life (i. e. biologists) of the day appeal, giving *theological* reasons for their beliefs similar to those of the astronomers and cosmologists.

Schleiermacher's account of the creation of animals capable of being conscious of their absolute dependence on God is famous for being a "developmental," or, "evolutionary" account.[86] Of course, Schleiermacher died decades before Charles Darwin published *On the Origin of Species.*[87] In consequence,

---

85 Kant, *Universal Natural History*, 82.

86 Schleiermacher often describes life's origins and change using *entwickeln* and its variants. For example: "*Können wir nun aber auch nach unserer erweiterten Weltkenntniss die Weltkörper mit allem auf ihnen entwikkelten Leben als einzelne Dinge ansehen, die nicht nothwendig alle gleichzeitig entstanden sind* [...]." *GL* §38.1; *The Christian Faith*, 146.

87 *On the Origins of Species* was first published in 1859.

Schleiermacher's "evolutionary" account is not obviously connected with Darwin's evolutionary account. But Charles Darwin was not the first to propose the evolution of new forms of life. More than one account of the evolution of new animal kinds preceded him.[88]

Moreover, the notion of the extinction of older forms of life, so essential a feature of Charles Darwin's account, had a longer pedigree still, though it was, at the same time, perhaps more controversial.[89] Even though Schleiermacher died twenty five years before *The Origin of Species*, the major requisite scientific notions needed to form a genuinely modern account of evolution were, as we will see below, in the air before Charles Darwin proposed natural and sexual selection as the mechanisms of biological evolution.

What specifies Schleiermacher's use of "evolution" as more than a pun is his use of the major notions of modern biological evolution: new species and the extinction of old ones. As he says in the context of his doctrine of creation, "we know fairly certainly of our earth that there have been species on it which are no longer extant and that the present species have not always existed; so that our proposition [regarding divine preservation] must be stretched to embrace these also."[90] This claim makes explicit that Schleiermacher has biological evolution, in a robust sense, in mind, and that he intends to propose or subscribe to more than a vaguely "developmental" account of creation or of human origins.

Schleiermacher's claim that new *species* have come to be distinguishes it from many pre-Darwinian views. Notions of change and variation *within* natural kinds date back to the ancients.[91] That is, many held that species vary internally, but, nevertheless, *as kinds* remain fixed. Without the additional notion of evolution of natural kinds themselves, this kind of change is unremarkable. Schleiermacher's view is more radical. It implies, most importantly, that he does not merely subscribe to the internal variation of species. As for the appeals of the tradition to the *Genesis* narrative for the fixity of species, Schleiermacher has his doubts. "Even what is said about the naming of animals," he says, "leaves us quite uncertain whether the designation had any regard to the relation of the kinds of animals to their species and of the species to the larger classes,

---

88 Darwin notes a number of precursor views in Appendix 1 of the *Origin*, "An Historical Sketch," including (but not limited to) Buffon, Lamarck, Goethe, and his grandfather Erasmus, with varying degrees of approval. Charles Darwin, *On the Origin of Species*, ed. William Bynum (London: Penguin [1859] 2009), 429–439.

89 Again, see Darwin's own remark on the subject in his "Historical Sketch," *Origin of Species*, 429.

90 *GL* §46, postscript; *The Christian Faith*, 175.

91 See, for example, Basil, *Hexaemeron*, VII, 2.

and if so, to what extent."[92] It is instead, for Schleiermacher, new species, new natural kinds, which emerge in the course of biological evolution, not merely variation within existing species.

However, for there to be species on this earth which "have not always existed" implies that some species came to be, but it does not yet specify *how* they came to be. They might, for example, have been created by special divine act. Or, instead, they could have developed entirely through natural processes. Schleiermacher's view is decidedly the latter. As he says:

> Now, with our increased knowledge of the world, we may indeed conceive the heavenly bodies and all the life developing upon them as particular things which have not all necessarily come into existence simultaneously; yet their successive origination must obviously also be conceived as the active continuance of formative forces which must be resident in finite existence.[93]

In Schleiermacher's view there was no special divine act which brought about the emergence of new natural kinds, in exactly the same way that there was no special divine act that formed new celestial bodies from old components, but instead, both developed wholly through natural processes, a power nascent in creation.

However, even if the history of life is an emergence of new natural kinds over time, it does not necessarily follow that Schleiermacher subscribes to yet another of the characteristic notions of modern evolution: that the origin of life itself was a natural process. That is, Schleiermacher might have held that only once life existed was it subject to the processes of biological evolution, but that the emergence of life in the first place was nevertheless a special divine act. Here too, however, Schleiermacher holds that *life itself* emerged through natural forces. To quote him at length:

> If, as suggested above, we strictly isolate the first creation, so that all things not absolutely primitive are regarded as part of the developing process of nature and thus brought under the conception of preservation, then the question whether creation occupied time, is answered in the negative. The distinction between a first and second creation, or an indirect and direct creation, always comes back in general to the evolving of the complex from the simple and of the organic from the elementary. But to acknowledge another creation here is either again to entirely abolish the difference between creation and preservation or to assume different kinds of matter devoid of inherent forces, which is surely meaningless. But even if in the case of creation we think first of matter (though we might equally well think of forces), then, from that point, living mobile being must have existed and undergone a con-

92 *GL* §61.2; *The Christian Faith*, 249.
93 *GL* §38.1; *The Christian Faith*, 146.

> tinuous development. Otherwise the creation of bare matter would have only been a preparation, that is an external material corresponding to the previously mentioned inner formal one. We must refer these definitions back to a time when men delighted in such abstractions because there was then no question of a dynamic aspect of nature.[94]

That is, the continuity of development is not merely from natural kind to natural kind but from primordial forces to full-fledged life. Life itself is a strictly natural development. Any distinction between two creations—one of matter, the other of life—either entirely abolishes the distinction between creation and preservation or implies that there is "matter devoid of inherent forces [...]." On the one hand, Schleiermacher defends the distinction between creation and preservation so as to prevent the confusion of the divine origin of time with an activity *in* time.[95] And, on the other, the notion of matter devoid of inherent forces is meaningless by definition for Schleiermacher—a claim that is explained further in Chapter 4. But if there was no separate act of divine creation in time, and indeed, no such thing as bare matter to add forces to, then, from the creation of the world on, the world must have implicitly contained the forces that would develop into life—"living mobile being"—and that development was in total continuity with the elementary stuff of which life is made. The development from non-life to life was "continuous."

Schleiermacher's belief that complex life evolved from simpler forms of life, and ultimately from non-life, all through the powers and forces inherent in the universe, is additionally supported by his nonchalant subscription to the likelihood of extraterrestrial life. "Now we to whom the majority of the heavenly bodies are known," Schleiermacher says, "satisfy this longing by the familiar supposition that most or all of these are filled with animated beings of varying grades."[96] This corroborates his above positions and makes clear how really *universal* he (and many of the evolutionists of the day) thought the operations of nature which bring about complex life from its simplest origins were. Put differently, one could have high expectations of life evolving on other planets only if one already assumed that the composition of the universe was relatively uniform and that the processes by which life emerged and developed were equally so. Only given the naturalness of life itself could one even possibly, "become fully

---

94 *GL* §41.2; *The Christian Faith*, 154–155.

95 "If the conception of Creation is to be further developed, the origin of the world must, indeed, be traced entirely to the divine activity, but not in such a way that this activity is thought of as resembling human activity; and the origin of the world must be represented as the event in time which conditions all change, but not so as to make the divine activity itself a temporal activity." *GL* §41; *The Christian Faith*, 152.

96 *GL* §42.1; *The Christian Faith*, 157.

persuaded that on every one of the planets organic life is developed up to the level of rationality [...]."[97]

Schleiermacher's subscription to the likelihood of extraterrestrial life is not to be taken for granted, but neither was it particularly rare. Indeed, the topic was widely debated in the late eighteenth and early nineteenth century.[98] Not coincidentally, both Kant in the *Universal Natural History*, and Herschel in various works, had argued for the inhabitance of many, if not most, celestial bodies. Herschel even argued that the surface of the sun was a likely habitat for beings suitably adapted.[99] Many of the same principles of universal development which Kant and Herschel applied to celestial bodies were also applied to animate bodies. The likelihood of extraterrestrial life was an obvious consequence of these principles. Both stars and life developed across the cosmos through powers and forces inherent in the world.

Since Schleiermacher subscribed to 1) the evolution of new forms of life, including humans, 2) from natural forces or processes, and 3) that such processes were so universal that the development of extra-terrestrial life could easily be expected, candidates for influence can be evaluated on the basis of how well they adhere to the same criteria. So, who are the candidates?

Though Charles Darwin is not a possible source of Schleiermacher's belief in the evolution of life, Charles's grandfather, Erasmus Darwin, is. Though Erasmus Darwin (1731–1802) did not propose the particular mechanism of natural selection, as Charles later would, what he did supply was a truly modern evolutionary account of species in the sense described above. That is, he proposed that life emerged naturally, apart from supernatural intervention, from non-living elements and forces, and that it developed from primordial states into all the historical and presently extant species.

In *The Temple of Nature, or, The Origin of Society*, Erasmus Darwin put his theory to verse. Here in poetic form, supported by explanatory notes, he outlines a remarkably modern version of the evolution of life—a doctrine of evolution which completely coheres with Schleiermacher's. To quote Darwin at length:

> Organic life beneath the shoreless waves
> Was born and nurs'd in Ocean's pearly caves;
> First forms minute unseen by spheric glass,
> Move on the mud, or pierce the watery mass;

---

97 *GL* §164.2; *The Christian Faith*, 724.

98 See Michael J. Crowe, *The Extra-Terrestrial Life Debate, 1750–1900* (Mineola, NY: Dover, [1986] 1999).

99 William Herschel, "On the Nature and Construction of the Sun and Fixed Stars," 63.

These, as successive generations bloom,
New powers acquire, and larger limbs assume;
Whence countless groups of vegetation spring,
And breathing realms of fin, and feet, and wing.
Thus the tall Oak, the giant of the wood,
Which bears Britannia's thunders on the flood;
The Whale, unmeasured monster of the main,
The lordly Lion, monarch of the plain,
The Eagle soaring in the realms of air,
Whose eye undazzled drinks the solar glare,
Imperious man, who rules the bestial crowd,
Of language, reason, and reflection proud,
With brow erect who scorns this earthy sod,
And styles himself the image of his God;
Arose from rudiments of form and sense,
An embryonic point, or microscopic ens![100]

Lest the classical style of Darwin's poetry obscure the novelty of the theories it promotes, more than one notable point should be made plain. The first is that these diverse and extraordinary animals, distinct as they are in bodies, powers, and native habitat, all spring from the same microscopic predecessors. Humans are no exception, but share in that common derivation. Second, vegetable life is included in this continuity as well. Both oak and lion come from one and the same ancestor. There is a unity of origins for all natural kinds, and no suggestion of a plurality of *fundamental* forms or species, of which later developments are mere variations upon or within. Third, and finally—and what is most important for comparison with Schleiermacher—life itself spontaneously arises from non-life. That is, no special divine act is required to bring forth life. Life arises solely through the powers and forces inherent within nature itself.

Darwin explains this doctrine in his notes under the heading of the "Spontaneous Vitality of Microscopic Animals."[101] Like Kant above, he feels compelled to begin by addressing objections— or "prejudices," as he calls them—to his theory. And also, like Kant, Darwin seems to regard *theological* objections as the most pertinent, and his own theory as not simply a better account of life, but a more adequate theological account as well. To quote him again at length:

> From the misconception of the ignorant or superstitious, it has been thought somewhat profane to speak in favour of spontaneous vital production, as if it contradicted holy writ; which ways, that God created animals and vegetables. They do not recollect that

100 Erasmus Darwin, *The Temple of Nature, or, the Origin of Society: A Poem with Philosophical Notes* (London: 1803), 6.
101 Darwin, *The Temple of Nature*, 36.

> God created all things which exist, and that these have been from the beginning in a perpetual state of improvement; which appears from the globe itself, as well as from the animals and vegetables, which possess it. And, lastly, that there is more dignity in our idea of the supreme author of all things, when we conceive him to be the cause of causes, than the cause simply of the events, which we see; if there can be any difference in infinity of power![102]

Just as Kant argued for the sufficiency and so continuity of the natural order in the development of the celestial realm, so Darwin argued for the same in regard to the terrestrial. And, while both appealed to more than the adequacy of their view of divine action and the inadequacy of their opponents', they each began with God.

This approach to biological evolution was not restricted to Erasmus Darwin alone. Jean Baptiste Lamarck (1744–1829) was Darwin's (and Schleiermacher's) contemporary, and an important authority on evolution until Charles Darwin's *Origin of Species*. Lamarck, like Erasmus Darwin, argued against the fixity of species and for the emergence of new natural kinds through equally natural processes. This view, nearly unique in his time,[103] was that all the extant species were not, as Lamarck so simply and memorably put it, "as old as nature."[104] Moreover, Lamarck did not restrict this change to life only once it existed, but included life's *origins* as well as its variations in the process of natural development. In fact, he made this point the subject of his first two principles:

> (1) That all the organized bodies of our earth are true productions of nature, wrought successively through long periods of time.
>
> (2) That in her procedure, nature began and still begins by fashioning the simplest of organized bodies, and that it is these alone which she fashions immediately, that is to say, only the rudiments of organization indicated in the term *spontaneous generation.*[105]

That is, not only are *all* living things on earth made by *nature*, but more complex forms of life are developed through the most simple. In the relevant respects, Schleiermacher's view of the evolution of life was equally coherent with Lamarck's view as it was with Erasmus Darwin's.

102 Darwin, *The Temple of Nature*, 36.
103 Jean Baptiste Lamarck, *Zoological Philosophy*, trans. Hugh Elliot (London: Macmillan, [1809] 1914), xxii.
104 Lamarck, *Zoological Philosophy*, 36.
105 Lamarck, *Zoological Philosophy*, 40.

Lamarck, again like Kant *and* Darwin, defended his view against theological objections with his own theological arguments. With Kant, he began with a rhetorical question from his hypothetical objector:

> Will anyone, it may be asked, venture to carry his love of system so far as to say that nature has created single-handed that astonishing diversity of powers, artifice, cunning, foresight, patience, and skill, of which we find so many examples among animals? Is not what we see in the single class of insects far more than enough to convince us that nature cannot herself produce so many wonders; and to compel the obstinate philosopher to recognize that the will of the Supreme Author of all things must be here invoked, and could alone suffice for bringing into existence so many wonderful things?[106]

Combining explanatory lack with the charge of impious curiosity, Lamarck's hypothetical objector seems to have God on his side. But Lamarck turns the tables. "No doubt he would be a bold man, or rather a complete lunatic," he says, "who should propose to set limits to the power of the first Author of all things; but for this very reason, no one can venture to deny that this infinite power may have willed what nature herself shows us it willed."[107] Instead, if natural causes should be found for natural things, Lamarck asks:

> Shall I admire the greatness of the power of this first cause of everything any the less if it has pleased him that things should be so, than if his will by separate acts had occupied itself and still continued to occupy itself with the details of all the special creations, variation, developments, destructions and renewals, in short, with all the mutations which take place at large among existing things?[108]

No, those are rather his objector's burdens. Lamarck is content to see that nature produces nature's own wonders by nature's own powers—powers endowed to nature by God.

In fact, the most striking feature of both Darwin's and Lamarck's accounts is their commitment to the sufficiency of nature in explaining the origin and development of life, and the non-fixity of natural kinds necessary to secure it. In spite of their very different explanations for how this change comes about, both were committed to the developmental hypothesis—a hypothesis which both supported, and which was supported by, a distinct account of divine action.

106 Lamarck, *Zoological Philosophy*, 40.
107 Lamarck, *Zoological Philosophy*, 41.
108 Lamarck, *Zoological Philosophy*, 41.

Stars, Life, and Natural Development

That the above are largely theological disputes might suggest that Darwin and Lamarck were primarily debating a theological point for its own sake. This would, however, be misleading. Instead, their responses to objections were just that. Certain theological commitments, antithetical to their own theories, became impediments to be discarded or overcome, and to be replaced by different theological commitments which both supported, and were supported by, their specific scientific theories in turn.

Notice the form these arguments took. As Lamarck made clear, the virtues of his position are its ability to explain the natural world.[109] Just as with Kant, Herschel, and Laplace, Darwin and Lamarck too found support for their explanatory principles in the phenomena. As the observation of nebulae supported certain hypotheses about the formation of stars, so observation of living things supported specific hypotheses about the development of new forms of life. Darwin's and Lamarck's developmental principles, like the astronomers' above, functioned only as *hypothetical* principles in a technical sense: principles supported by their explanatory power, but not by higher principles. In fact, these scientists' burden was just the opposite: that the highest principles to which one might appeal—specifically, theological commitments—*contradicted* their explanatory hypotheses. It was to remove these obstacles that the above natural scientists appealed to divine things.

This insight begins to unravel more than one mystery. In the most immediate sense it clarifies why these natural scientists thought theological questions were pertinent to discussions of stars and life. If their version of the natural world was incompatible with certain assumed theological commitments, either their natural philosophy or their theology needed to be modified. Rather than wholesale abandonment, the above thinkers sought to open up theological options that others had considered shut, rather than discard theological commitments altogether.

Observation and mediating principles supported this revision of higher principles. Mediating principles are those which explain observations by providing them with a structure of intelligibility. A favorite principle of this sort for all of the above natural scientists was, as we saw, the principle of analogous development. It was analogy which suggested that disparate momentary celestial observations were, in fact, different stages of a common development, and that mi-

---

109 Much of Lamarck's remaining efforts of the *Zoological Philosophy* are devoted to advertising the strengths of his theory by its explanatory power.

crobes growing in a dish were much like the very origins of life on earth. It was only through the principle of development that inferences could be made from what the world is like at the present to what the world was like in the past. It was only through such unifying principles that a *developmental* universe—celestial and/or terrestrial—could be plausibly proposed. And, as that principle was itself supported by its explanatory power, so other principles from which development follows were in turn supported by the explanatory power of the developmental hypothesis.

The above thinkers appealed to the success of their explanations of natural phenomena through the developmental hypothesis as reason in turn to find their view of divine action more plausible. If God, in fact, acts only through the powers and forces inherent in the natural world, and everything that happens in nature is itself a product of nature, that would very much support the constancy of the world which development partially relies upon. At the same time, if through such constant means, change were in-built, development itself would be naturally ordered and divinely ordained. In this way, observational findings of natural scientists supported explanatory principles which implied that the universe operated in very general ways and not others, and which, in the end, supported a particular view of divine action in relation to that universe over and against other views of divine action.

In the following chapter we will see how one of the most famous scientific, philosophical, and theological debates of the eighteenth century hinged on this very appeal to divine action. Since the argument goes in both directions, it was not only possible to support certain views of divine action by appeal to what they explain in nature, but also to explain away certain views of nature by appeal to divine action as a premise. Schleiermacher, following this strategy as well, appeals from first principles to an account of divine action *at the same time* that he appeals to the natural science of the day for explanatory support from below. That is, Schleiermacher argues in the same manner that natural scientists did, only starting in reverse, to a unified account of reality that coordinates first principles with mediating principles, and principles of all kinds with content.

But, for now, one last mystery remains: Schleiermacher on the comparison of Christ and nebulae. Having seen what was at stake in the debates over nebular formation as an instance of development, and how natural scientists of the day appealed to both natural phenomena and divine things in support of their theories, we are now in a position to make sense of Schleiermacher's comparison.

It was not ordinariness that connected the appearance of the redeemer to the formation of nebulae. The former is unique; only the latter common. Instead, it was the *natural development* of both which the comparison regarded. Just as new things (stars) arise naturally from the universe, so too Christ's incarnation is a natural phenomenon. Like life itself, the conditions for the incarnation were contained in the universe even as, as also with life generally, Christ's spiritual existence was *relatively* novel. His arrival was new only in the sense that the formation of nebulae and stars had not yet happened before they happened. And yet, like life and stars, the appearance of the redeemer in the world was always *potential* in the strong sense that the sufficient conditions for that appearance were already inherent in the world. As stars are contained virtually in nebulae, and nebulae are contained virtually in the celestial stuff of which they are made; so too, the redeemer was contained virtually in the world and, in particular, in human nature from the beginning.[110]

Moreover, the process of arriving at such a conclusion proceeds in the same way for stars, life, and Christ. In each case, principles are proposed as hypotheses to explain phenomena. These principles are supported by appeals to like occurrences which those principles make intelligible. And, in this case, one and the same developmental principle explains *both* nebulae *and* the appearance of the Redeemer. Schleiermacher's comparison is not a mere metaphor. It is more than that. Schleiermacher's comparison is, in fact, an appeal to the naturalness of Christ's appearance *which repeats the strategy employed by the natural scientists of his day* who used the explanatory power of mediating principles as warrant to modify (or specify) higher principles. Crucially, the more that can be explained under these mediating principles, the greater their support. And so, with "the coming of Christ" Schleiermacher offered one more occurrence which could be explained under the developmental hypothesis. In so doing, he both gave support to, and borrowed support from, the mediating principles of the natural scientists of his day—mediating principles which all parties brought to bear on questions of divine action. In other words, Schleiermacher's comparison of the incarnation to nebulae was not merely a metaphor, but evidence.

---

110 See *GL* §13.1; *The Christian Faith*, 62–64.

Chapter 3 Divine Wisdom and the Order of the World: Leibniz and Schleiermacher on the Perfection of Nature

> "Let me tell you then why the creator made this world of generation. He was good, and the good can never have any jealousy of anything. And being free from jealousy he desired that all things should be as like himself as they could be. This is in the truest sense the origin of creation and of the world, as we shall do well in believing on the testimony of wise men: God desired that all things should be good and nothing bad, so far as this was attainable."
>
> - Plato, *Timaeus*

Early on in his dispute-by-correspondence with Isaac Newton's defender Samuel Clarke,[111] Gottfried Leibniz addressed a problem with Newton's natural philosophy. If gravity was the sort of force which could act at a distance without a mechanical medium, and if there was no countervailing force in the universe, then eventually, as Newton himself concluded, the pull that planets and passing comets exert on one another would destabilize the orbits of their neighbors and the fixed stars of the heavens would collapse in upon one another. The accumulating effects would be slow, but over a long enough period of time that slight pull would result in the dissolution of the solar system, "till this system wants a reformation."[112] The natural effects of gravity would be catastrophe, and such a consequence would surely contradict God's providence. However, while such an end result was incompatible with the divine government of the world, Newton saw no reason why God could not simply set the world right again, why God could not easily rectify the disturbance of a comet here or there, and replace the stars in their proper station.

Leibniz was unconvinced. He found that Newton's solution, that God would simply reset the world from time to time, presented its own grave difficulties. These difficulties turned a dispute, which originally arose over occult explanations, action at a distance, and the eventual effects of the force of gravity, into

111 H. G. Alexander, ed. *The Leibniz-Clarke Correspondence*, (Manchester: Manchester University Press, 1956). The translation I use throughout is Clarke's own 1717 rendering.
112 Isaac Newton, *Opticks* (1704), Query 31

DOI 10.1515/9783110542301-003

a theological argument. It was not so much that Newton could offer no explanation of how God might maintain the world in spite of gravity—his *deus ex machina* solution was, after all, a solution. Rather, Leibniz did not so much object to Newton offering a divine solution to a cosmic problem, but objected more basically to Newton's having admitted a problem in the first place. Leibniz countered that, given God's wisdom, one would not expect the kind of divine action that Newton's teaching seemed to demand. And, as a result, since Newton's natural philosophy seemed to *require* special divine action from time to time, one would not expect the natural philosophy Newton proposed to be true.

Not only Newton's theology, but also his science, hung in the balance. Both parties agreed that God ordained a great end for the world, an end worthy of the divine goodness. And both agreed that by God's omnipotence that end would surely be attained. But the issue of divine agency was thrown into stark outline since, unlike all other agents, God not only *used* means for an intended end, but also *created* those means in the first place. Creation was teleologically ordered from its beginning. Since Newton's natural philosophy impugned the divine wisdom by suggesting a discrepancy between God's means and ends, Leibniz concluded that Newton's science must be false.

As it turned out, Newton was wrong about the destabilizing effects of gravity, and the scientific worries of both parties were for naught.[113] Nevertheless, the theological dispute which arose over the intersection of gravity and providence remained important. What Leibniz argued for was a critique of divine supplementation or emendation of the natural order on theological grounds, and his argument transcended his issues with the details of Newton's natural philosophy. Leibniz's was a providential critique of miracles which not only relied on divine power, but also divine goodness. Under the demands of the principle of sufficient reason (see below), Leibniz challenged Clarke to make certain all explanations terminated in God *and* to ground those answers in something other than divine caprice. In short, Leibniz claimed that God's power was insufficient reason apart from God's wisdom. And although Leibniz's argument was couched in his own characteristic terms, his argument can easily be abstracted from its original context and redeployed in different language.

Friedrich Schleiermacher does exactly this in *The Christian Faith*. He and Leibniz share the same premises. They share the same conclusions. There is a di-

113 The discovery that gravitational forces tended on the whole to counteract one another, and so did not degrade the world-order over time, saw the resolution of the dispute. Newton's celestial mechanics became compatible with Leibniz's argument from divine wisdom. Only with that issue resolved was Newton's natural philosophy more widely adopted. Alexander, *The Leibniz-Clarke Correspondence*, xvii, n 1.

rect historical link between the two. And so it is mostly likely that Schleiermacher simply adopted Leibniz's argument for his own doctrinal purposes.

This connection is illuminating. In Leibniz's thought, as in Schleiermacher's, three things are all found together: one, a direct interest in specific scientific theories; two, a general program for the study of the natural world; and three, concrete theological commitments. Just as for Schleiermacher, Leibniz's doctrine of God supports a teleologically ordered world, which in turn supports a view of nature as developing solely through the powers inherent in it. His commitment to the continuity of the natural order in turn supports his efforts to explain natural events solely though natural causes, which finally serves as reason to prefer particular scientific theories over others. Those theories which contradict the continuity of the natural order are correspondingly less plausible. All this follows from a series of interconnected commitments including a particular doctrine of God in exactly the same way that the natural science of the day coordinated principles and content—only this time descending from first principles rather than ascending from phenomena. In *The Leibniz-Clarke Correspondence*, Leibniz uses this whole chain of mutually supporting beliefs explicitly to settle a theological dispute *and* a natural scientific dispute *simultaneously.*[114] If Schleiermacher's position derives at least in part from Leibniz's, that makes Schleiermacher's interconnected use of theological commitments and natural scientific commitments all the more apparent and understandable. This chapter will begin to examine a series of arguments that collectively show that Schleiermacher's account of God and divine action was meant to coordinate theology and science in just this way, not divide them.

The content of Leibniz's and Schleiermacher's claims, and the arguments they give in support of those claims, are crucial. For instance, both positions include an assumed determinate teleology. Clarke also shared this premise and so offered no challenge to Leibniz on this basis. The frequent discussion of the aptness of God's means to God's ends obviously implies that there is, in fact, an end of God's activity. Schleiermacher is even more explicit that the *telos* of the world is revealed in the God-consciousness of the redeemed as the redemption accomplished by Jesus of Nazareth.[115] For him, Christianity *is* teleological moral monotheism.[116] But if both Schleiermacher and Leibniz are arguing *from* divine purpose to a view of the world, then they are definitely not attempting to argue

114 For a detailed exploration of the importance of final causes in general, and theology in particular, for Leibniz's physics, see Daniel Garber, *Leibniz: Body, Substance, Monad* (Oxford: Oxford University Press, 2009), especially Chapter 6—"Divine Wisdom and Final Causes."
115 *GL* §164; *The Christian Faith*, 723.
116 *GL* §11; *The Christian Faith*, 52.

from the world *to* divine purpose in the style of someone like William Paley.[117] The direction of argument is exactly the opposite of what later generations will come to think of as natural theology.[118]

At the same time, neither Schleiermacher nor Leibniz thinks that knowledge of the world is irrelevant for knowledge of God. On the contrary, both seem to think it vital. It is this point which will be borne out at some length in Chapter 6. For Leibniz, the *reason* that God makes the world one way and not another is the divine wisdom and goodness. If God were indifferently disposed to the world being one way and not another then God would not act for the *best*, that is, for the good. And although Schleiermacher rejects Leibniz's metaphysics of possible worlds (and so rejects the notion that this world is best if "best" implies that there are other, genuinely possible, worse possible worlds), the coordination of ends and means is still crucial for him. As we will see in Chapter 6, for Schleiermacher it is the absolute unity of God's nature with God's activity, and of God's ends and God's means, that makes *the world* the revelation of God. The natural world is ordered to God's ordained end, and as God is not indifferent to the means, Christians should not be indifferent either. Only if the natural world is the revelation of God, Schleiermacher argues, with Christ as an integral feature of that world, can God be called wise and the world be known as good.

For both Leibniz and Schleiermacher, the argument from divine wisdom is an argument against supernatural supplementation of the natural order on the basis of divine perfection. The wiser God is, and the more the natural world expresses the wisdom of God, the less we should expect absolute miracles (understood as events in the nature system that exceed the power not only of particular created things, but nature as a whole—see Chapter 4). The perfection of the natural order renders divine supplementation superfluous. As a result—a conclusion which might initially strike some as odd—the less God *discretely* acts in history, the greater the corresponding perfection of God's creation. The more perfect the art, the more perfect the artist. Only if the natural world never needs, and so never includes, discrete supernatural acts, is the world perfect and God absolute-

---

117 I specifically mean that Leibniz's argument in *The Leibniz-Clarke Correspondence* is premised on an assumption that divine action is ordered to an end. The arguments for the existence of God Leibniz pursued elsewhere are beyond my present claim and interest. See Paley's 1802 work *Natural Theology*, Matthew D. Eddy and David Knight, eds. (Oxford: Oxford University Press, 2008).

118 For an argument for the importance of this Christological commitment see Bruce L. McCormack, "What Has Basel to Do with Berlin? Continuities in the Theologies of Barth and Schleiermacher" in *Orthodox and Modern: Studies in the Theology of Karl Barth* (Grand Rapids: Baker, 2008).

ly wise. Of course, this is only the argument in its nascent form. For a fuller understanding we must turn again to Leibniz's dispute with Clarke.

Leibniz Against Clarke on the Wisely Ordered World-Clock

Leibniz's dispute with Clarke, defender of Newton's natural philosophy, began with a letter to Caroline, Princess of Wales.[119] In that letter Leibniz charged Newton's thought with many faults including, "a very odd opinion concerning the work of God."[120] As he told the Princess:

> According to their [Isaac Newton and his followers] doctrine, God Almighty wants to wind up the watch from time to time: otherwise it would cease to move. He had not, it seems, sufficient foresight to make it a perpetual motion. Nay, the machine of God's making, is so imperfect, according to these gentlemen; that he is obliged to clean it now and then by an extraordinary concourse, and even to mend it, as a clockmaker mends his work; who must consequently be so much the more unskilled a workman, as he is oftener obliged to mend his work and set it right.[121]

From the very first volley, Leibniz's criticism of Newton's natural philosophy hinged on the question of divine agency, in particular God's wisdom.

Clarke, however, answered in reply that Leibniz's diagnosis of the dis-analogy with the watchmaker—that, unlike human watchmakers, God need never clean, wind, or mend—was wrong. God's special dignity did not so much come from the fact that God made a better watch than all other watchmakers, and so never mended his creation:

> But with regard to God the case is quite different; because he not only composes or puts things together, but is himself the author and continual preserver of their original forces or moving powers: and consequently 'tis not a diminution, but the true glory of his workmanship that nothing is done without his continual government and inspection.[122]

Moreover, Clarke was unflinching in his own diagnosis of Leibniz's position:

> The notion of the world's being a great machine, going on without the interposition of God, as a clock continues to go without the assistance of a clockmaker, is the notion of materi-

---

119 Leibniz's First Paper in Alexander, *The Leibniz-Clarke Correspondence*, 11.
120 Alexander, *The Leibniz-Clarke Correspondence*, 11.
121 Alexander, *The Leibniz-Clarke Correspondence*, 11–12.
122 Clarke's First Reply in Alexander *The Leibniz-Clarke Correspondence*, 14.

> alism and fate, and tends, (under pretense of making God a *supra-mundane intelligence,*) to exclude providence and God's government in reality out of the world.[123]

If the metaphor of a king were substituted in place of a watchmaker, charged Clarke, Leibniz's God would not be so impressive. His God would merely be the nominal king of a nominal kingdom,[124] a sovereign in name only.

Appearing to sense the need for care, Leibniz began his second paper in response to Clarke slowly, laying out his *principle of sufficient reason*, "that nothing happens without a reason why it should be so rather than otherwise,"[125] before setting out to counter Clarke. Taking up the question of the divine watchmaker once more, he argued that:

> The true and principal reason why we commend a machine, is rather grounded upon the effects of the machine, than its cause. We don't enquire so much about the power of the artist, as we do about his skill in his workmanship. And therefore the reason alleged by the author [Clarke] for extolling the machine of God's making, grounded upon his having made it entirely, without wanting any materials to make it of; that reason, I say, is insufficient. 'Tis a mere shift the author is forced to have recourse to: and the reason why God exceeds any other artist is not only because he makes the whole, whereas all other artists must have matter to work upon. This excellency in God, would be only on account of power. But God's excellency arises also from another cause, viz. wisdom: whereby his machine lasts longer, and moves more regularly, than those of any artist whatsoever.[126]

According to Leibniz, Clarke's case was weak because he only appealed to divine power, and not also to divine wisdom. Because his case did not take account of divine wisdom, Clarke could not give an account of the sufficient reason of God's action.

What a sufficient reason would be in this case—that is, what exact appeal to wisdom is needed—is not quite clear. It is not obvious, at least, that the world's *duration* is proportional to God's wisdom, so that the longest lasting world necessarily exhibits the greatest wisdom of its maker. But Leibniz did not seem to have meant to give such a reason, only to state that there must be one. His emphasis was, rather, on the formal point that the necessary condition of the world's having been made by God is *not yet* a sufficient condition for it having been made by God *wisely*. As he said to Clarke, "[H]e who will be pleased with

123 Alexander, *The Leibniz-Clarke Correspondence*, 14.
124 Alexander, *The Leibniz-Clarke Correspondence*, 14.
125 Leibniz's Second Paper in Alexander, *The Leibniz-Clarke Correspondence*, 16.
126 Alexander, *The Leibniz-Clarke Correspondence*, 17–18.

God's workmanship, cannot be so, without some other reason than that which the author has here alleged."[127]

For a standard of wisdom, Leibniz initially relied on his shared assumptions with Clarke about God's good purposes, however generic. In his watch analogy he mentioned the watch needs to work right[128] and that the watchmaker would not be considered skilled, even if able to create the parts of the watch from nothing, "unless the workman had also received the gift of putting them well together."[129] The language of *proper* function, and of being assembled *well* is, of course, the language of final causes, the language of purpose. It is without these final causes that, Leibniz claimed, no sufficient reason can be given for the world being thus and not otherwise. "The bare production of every thing, would indeed show the *power* of God," Leibniz explained, "but it would not sufficiently show his *wisdom*."[130]

Initially, Leibniz offered no concrete purposes. However, as his dispute with Clarke intensified, he became more explicit. His contention, he revealed later on in the dispute, was precisely over the *providential* decree of God.[131] It is a happy decree for Christian believers,[132] implying the love of God,[133] in, "knowing he [God] does everything for the best; and not only for the greatest good in general, but also for the greatest particular good of those who love him."[134] What Leibniz initially spoke of only in generic terms was not only a generic good, but a highly determinate good with universal implications.

---

127 Alexander, *The Leibniz-Clarke Correspondence*, 18.
128 Alexander, *The Leibniz-Clarke Correspondence*, 18.
129 Alexander, *The Leibniz-Clarke Correspondence*, 18.
130 Alexander, *The Leibniz-Clarke Correspondence*, 18.
131 Leibniz's Fifth Paper in Alexander, *The Leibniz-Clarke Correspondence*, 58.
132 Leibniz specifically calls this a "*fatum christianum*," a "Christian fate," distinguished from other species of fate by the happiness it brings owing to the goodness it recognizes. See Alexander, *The Leibniz-Clarke Correspondence*, 58. See also Leibniz's remarks in the *Theodicy* where he compares his own view to that of the Stoics and Epicureans: "It is no small thing to be content with God and with the universe, not to fear what destiny has in store for us, nor to complain of what befalls us. Acquaintance with true principles gives us this advantage, quite other than that the Stoics and Epicureans derived from their philosophy. There is as much difference between true morality and theirs as there is between joy and patience; for their tranquility was founded on necessity, while ours must rest upon the perfection and beauty of things, upon our own happiness." Leibniz, *Theodicy: Essays on the Goodness of God, the Freedom of Man, and the Origin of Evil*, ed. Austin Farrer. trans. E. M. Huggard. (La Salle, IL: Open Court, [1951] 1985), §254; 283.
133 Alexander, *The Leibniz-Clarke Correspondence*, 58.
134 Alexander, *The Leibniz-Clarke Correspondence*, 58.

By Leibniz's fifth paper, it was evident that it was this very *particularity* of ends that he sought with his use of the principle of sufficient reason in the first place, and which he felt that principle demanded. "For a man never acts without a sufficient reason to act," he said, "when he has not also a sufficient reason to act in a particular manner; every action being individual and not general, nor abstract from its circumstances, but always needing some particular way of being put in execution."[135] To act without an end is to act without sufficient reason. Moreover, to act with an end is always to act with a *determinate* end (ends being necessarily one thing and not another), and so a sufficient reason of action will always be a particular end. Further, a determinate end thus renders the question of means determinate as well—every means being either better or worse in relation *to* an end. "Wherefore," as Leibniz said,

> when there is a sufficient reason to do any particular thing, there is also sufficient reason to do it in a certain particular manner; and consequently, several manners of doing it are not indifferent. As often as a man has sufficient reason for a single action, he also has sufficient reason for all its requisites.[136]

From a sufficient reason not only is the particularity of that end implied and so the end rendered determinate, but also—and this is key—the means to that end, and the *entire* chain of proximate ends *and* their means, all linked in the same determinate way to the ultimate end in the series. Because both Leibniz and Clarke agreed that there was at least one divine purpose for the world, and so agreed that the world was indeed created for an end, Leibniz's accusation hinged on the *aptness* of the world to that end. Clarke certainly agreed that the world had a purpose, but, according to Leibniz, he could not give a sufficient account of how the natural world and its purpose related. If the world is a means to some divine end, then God must have sufficient reason for the world to be a certain way.

Importantly for Leibniz, God's wisdom implied not only a purpose and a means but, "all its requisites."[137] That is, God's wisdom implies not only that God acts for an end, and not only that the means to that end actually accomplish what they are for, but also that those means are themselves so well-ordered that they are in need of nothing else. God's wisdom implies that God's means are all-sufficient to their ends. To use Leibniz's watch metaphor, a watch made by God would not only be for a good end (in this case, telling time), but also be so well

135 Alexander, *The Leibniz-Clarke Correspondence*, 60.
136 Alexander, *The Leibniz-Clarke Correspondence*, 60.
137 Alexander, *The Leibniz-Clarke Correspondence*, 60.

made so as to accomplish this end inevitably (that is, it would *actually* tell time well), and—this is key—have implicit within itself, from the beginning, the power to do so. Its gears would turn so reliably, its hands would swing so precisely, that it would never need to be cleaned, wound, or repaired. And it would never need to be supplemented or emended precisely because the perfection of such a watch would lie in its *inherent* ability to actualize the end for which it was made.[138] Any supplement would imply a need, and so the insufficiency, and so the imperfection of the watch. Any imperfection of the watch would further imply the imperfection of its maker. "No," said Leibniz, "God has foreseen every thing; he has provided a remedy for every thing before hand; there is in his works a harmony, a beauty, already pre-established."[139]

God, according to Leibniz, arranged the universe such that all that happens does so according to God's ends. "But God never takes a resolution about the ends," Leibniz said, "without at the same time resolving about the means, *and all the circumstances*. Nay I have shown in my *Theodicy*, that properly speaking, there is but one decree for the whole universe, whereby God resolved to bring it out of possibility into existence."[140] God's wisdom demands a good end and perfectly apt means to that end. But the divine wisdom also demands the creation of all the circumstances surrounding that end and those means. Thus, God's wisdom is expressed as a *unity* of ends, means, and circumstances: a single divine decree. But notice: if God's wise decree includes not only a resolution of the ends to be attained, but *also* the means *and* their circumstances, then it is an exhaustively determined decree. Once God resolves to bring this universe into existence, God resolves for all that happens to happen in one way, the way it actually happens, and no other. The exhaustive particularity of God's decree becomes the theological underpinning of Leibniz's theory of *conceptual containment*.

The determinate particularity of the decree is not only noteworthy in connection with Schleiermacher, as we shall see, but it is also needed to address anoth-

---

138 Rutherford makes the case that Leibniz's point is even stronger: that not only must things include powers such that they are actually the cause of their effects, but moreover that such powers are unintelligible apart from natures through which those inherent powers can be explained. As we will see in Chapter 4, Schleiermacher makes the same move. See Donald P. Rutherford, "Natures, Laws, and Miracles: The Roots of Leibniz's Critique of Occasionalism" in *Causation in Early Modern Philosophy* (University Park, PA: Pennsylvania State University Press, 1993), 147.

139 Leibniz's Second Paper, in Alexander, *The Leibniz-Clarke Correspondence*, 18.

140 Leibniz's Fifth Paper, in Alexander, *The Leibniz-Clarke Correspondence*, 78. Emphasis added.

er potential objection to Leibniz's argument. This objection is, *contra* Leibniz, that some things in the world are not in fact divinely determined, and so God's relation to them might be partially responsive or reactive. If God does not exhaustively determine all that is to occur, then God might still wisely decree both ends and means but *not* all their circumstances. And if God's wisdom does not determine circumstances, then God might indeed supplement or emend the world so as to mitigate or augment the activities of *other* agents. If that were the case, arguments from God's wisdom would have little bearing on consideration of discrete divine activity and the natural order.

Leibniz, however, refused to admit that God fails to determine the circumstances of things while only determining their ends, and the means to those ends, in the abstract. His refusal echoes back to his claim against Clarke that sufficient reason does not consist in power *alone*.[141] For, although Leibniz denied that power is sufficient without wisdom, he did not deny that sufficient reason includes power. On the contrary, power is still a necessary, but not sufficient, condition of divinity. And so he argued against Clarke that:

> A true providence of God requires a perfect foresight. But then it requires moreover, not only that he should have foreseen every thing; but also that he should have provided for every thing before-hand, with proper remedies: otherwise he must want either wisdom to foresee things, or power to provide against them.[142]

When God acts with sufficient reason, God not only has a sufficient reason to act in a particular way, but also has the power to do so.

Leibniz's account of divine willing was indispensable to his attack against Clarke. It served his argument by implying that any supplementary activity on God's part is always a *self*-supplementation. Without this move, Leibniz's argument is weak. God might always supplement or emend the activities of *other* agents, agents whose actions might fail to satisfy, or perhaps even contradict, the divine purpose. The order of nature could instead be seen in line with the functioning of the laws of human society. The laws of nature might be normative, but not determinative. Indeed, a hallmark of human laws is that they are sometimes broken by troublesome subjects or overlooked by merciful rulers. The will of the sovereign and the activities of his subjects could fail to coincide. Or, the good will of the sovereign and his established laws might be at odds. Leibniz's account denies these possibilities. All things fall within the sphere of divine sov-

141 Leibniz's Second Paper, in Alexander, *The Leibniz-Clarke Correspondence*, 18.
142 Alexander, *The Leibniz-Clarke Correspondence*, 19.

ereignty, a dominion that, on the whole, "cannot diminish in perfection."[143] Only if *all* activity falls within the sphere of divine activity can Leibniz's charge hold that absolute miracles imply that, "God bethinks himself again."[144]

Further, Leibniz's account of God's exhaustive determination of all things furnished him with a retort to Clarke's charge that his God is a merely nominal ruler. Not only did Leibniz insist that the world is sustained in being at every moment by God ("I never gave any occasion to doubt, but that God's conservation is an actual preservation and continuation of the beings, powers, orders, dispositions, and motions of all things"[145]), but also that it is the very sovereignty of God that implied *his own* position. As Leibniz explained:

> The comparison of a king, under whose reign everything should go without interposition, is by no means to the present purpose; since God preserves every thing continually, and nothing can subsist without him. His kingdom therefore is not a nominal one. 'Tis just as if one should say, that a king, who should have originally taken care to have his subjects so well educated, and should, by his care in providing for their subsistence, preserve them so well in their fitness for their several stations, and in their good affection towards him, as that he should have no occasion ever to be amending any thing; would be only a nominal king.[146]

The greater the perfection of the king and so the more his ends accord with his means and all their conditions, the less his interposition is needed. Ironically, it is actually the king who rules through occasional supplementation or emendation whose reign is more nominal. Some things require special action only when they fail to proceed by his will. There are two very different reasons why divine interposition might be impossible or unnecessary: on the one hand, if what happens in the world falls outside of the sphere of divine power; and, on the other, if all that happens in the world *perfectly coincides with the divine rule*. The latter view, not the former, was Leibniz's.

Given the understanding of all parties at that time, that Newtonian natural philosophy *required* occasional divine supplementation or emendation, Leibniz's criticism was a powerful argument against Newton's position. Leibniz's alternative view of the universe was, in contrast, ordered and elegant: "As for the motions of the celestial bodies, and even the formation of plants and animals;" said Leibniz, "there is nothing in them that looks like a miracle, except their beginning."[147]

---

143 Leibniz's Fourth Paper, in Alexander, *The Leibniz-Clarke Correspondence*, 42.
144 Leibniz's Second Paper, in Alexander, *The Leibniz-Clarke Correspondence*, 18.
145 Alexander, *The Leibniz-Clarke Correspondence*, 29.
146 Alexander, *The Leibniz-Clarke Correspondence*, 19–20.
147 Leibniz's Fifth Paper, Alexander, *The Leibniz-Clarke Correspondence*, 93.

Schleiermacher's Connection to Leibniz

While the burdens facing Newton's natural philosophy were no longer under debate in Schleiermacher's day, the basic question of whether or not the universe and everything in it developed solely according to powers inherent in the natural order was alive and well. Divine wisdom remained pertinent in both theological and scientific circles. As we saw in Chapter 2, Schleiermacher's natural scientist contemporaries appealed to divine wisdom in just the same way. Across several centuries, divine wisdom was used to lend plausibility to certain theological views *and* certain scientific views alike. And, as we also saw in Chapter 2, Schleiermacher himself not only held natural scientific positions that are best suited to a developmentally contiguous universe, but held views that *explicitly* follow from appeals to divine wisdom—appeals which themselves proceeded along the very lines of Leibniz's position.

I do not, however, mean to imply that Schleiermacher subscribed to the majority of Leibniz's philosophy, let alone the whole of it. Indeed, some key differences between Schleiermacher and Leibniz will feature prominently in Chapter 5.[148] No, all Schleiermacher needed to borrow from Leibniz's arguments was enough shared premises to support the claims at hand. The following section on the details of Schleiermacher's argument will show that both figures, in fact, shared the same relevant premises. And the implications of this chapter together with Chapter 4 reveal more than that. Unlike the major arguments for the general regularity of the world from divine wisdom in the theological tradition—whether they be Augustine's or Thomas', or even Leibniz's own arguments elsewhere—Leibniz's argument in *The Leibniz-Clarke Correspondance*, like Schleiermacher's argument in *The Christian Faith*, admits of *no* exceptions, no instances where genuine miracles occur within creation. The precise form Schleiermacher's argument takes is exactly parallel to Leibniz's in this most important respect, and, even then, it is not found in all of Leibniz's works. Leibniz is, accordingly, the most plausible source for this unusual version of a more traditional argument.

Schleiermacher had access to this argument through several channels. He was both taught Leibniz's philosophy (or, rather, philosophy in the Leibnizian tradition) by J. A. Eberhard, and later made a study of Leibniz's works on his

---

148 For an overview of both similarities and differences between Leibniz's and Schleiermacher's philosophical commitments see Manfred Frank, "Metaphysical Foundations: A Look at Schleiermacher's *Dialectic*," trans. Jacqueline Mariña and Christine Helmer, in *The Cambridge Companion to Friedrich Schleiermacher*, ed. Jacqueline Mariña (Cambridge: Cambridge University Press, 2005), 15–34.

own.[149] Eberhard's teaching began in 1787, and Schleiermacher began to study Leibniz's own work in 1797.[150] So much is generally known. But what is often underemphasized is the prominence of Eberhard, his strong Leibnizian pedigree, and Schleiermacher's enduring respect for him.

Eberhard was one of, if not *the*, premier proponent of the Leibniz-Wolff school in his day.[151] Eberhard's pedigree was impressive. He was a student of Christian Wolff himself[152] who was, of course, Leibniz's most accomplished protégé. If there was anyone who could be counted on to transmit the philosophy of Leibniz reliably, it was Eberhard. Moreover, Schleiermacher had enormous personal respect for him. Even as Schleiermacher was more and more influenced by Kant, he continued to express his desire to consult with Eberhard, and continued to admire his intellectual prowess.[153] All this exposure to Leibniz's thought came a decade prior to Schleiermacher's direct reading of Leibniz's own works.

However, a question does remain as to the *degree* of Leibnizian influence. For example, a familiar line in some accounts is that while Schleiermacher was friendly to (or at least familiar with) the Leibniz-Wolff school early in his career, his reading of Kant so undermined his confidence in that tradition that its later influence is ultimately much less important than other intellectual sources.[154] There is some truth in this. That Kant had *some* enduring influence on Schleiermacher is not in doubt. And because Kant's ascendency came at the expense of the Leibniz-Wolff school in general, and at the personal expense of Eberhard in particular, that might suggest that the Leibniz school's influence on Schleiermacher correspondingly declined. This is also, to some extent, likely. However, the degree of decline should be evaluated *a posteriori*, rather than as-

149 See Frederica Rowan trans., *The Life of Schleiermacher*, (London: Smith, Elder and Co., 1860) vol. 1, 66, 69; For a more in-depth account of Schleiermacher's interaction with Leibniz see Jacqueline Mariña, *Transformation of the Self in the Thought of Friedrich Schleiermacher* (Oxford: Oxford University Press, 2008).

150 Given this early date, and the late publication of many of Leibniz's works, it might reasonably be asked which of Leibniz's works Schleiermacher actually had access to and likely consulted. However, as Brucker's widely-consulted work reveals, *The Leibniz-Clarke Correspondence* would have been prominent source, among others, including the *Theodicy*. See Iacobi Bruckeri [Jacob Brucker], *Historia Critica Philosophiae* (Leipzig, 1744), vol.4, part 2, 366–367.

151 See Norman Adams, *Schleiermacher's Philosophy of Freedom and His Relation to the Leibnizian Tradition*, Dissertation for the Syracuse Philosophy Department, 1941.

152 Adams, *Schleiermacher's Philosophy of Freedom*, 7.

153 Adams, *Schleiermacher's Philosophy of Freedom*, 8–10.

154 Brandt, for instance, tends to downplay Leibniz's influence. See Brandt, *The Philosophy of Schleiermacher*.

sumed from the start. If Schleiermacher borrowed his argument against supernatural interposition from Leibniz, then that should be precisely the sort of new information which should make us reevaluate the extent to which Schleiermacher discarded Leibniz.

Finally, lest the issue of sources or influences distract unnecessarily, I ask the skeptical reader to take such claims as conjectural, and to regard their main purpose as elucidatory. Whether or not parallels between arguments in Schleiermacher and Leibniz suggest that Leibniz is their source—though I am convinced they do—, Leibniz's argument nevertheless explains Schleiermacher's account by spelling out the same argument more clearly and completely. Influence aside, Leibniz's account greatly illuminates Schleiermacher's.

Schleiermacher on the Wisely Ordered System of Nature

All of Leibniz's major moves in his argument against miracles are repeated in *The Christian Faith*. First, Schleiermacher insists, like Leibniz, that divine power is insufficient without a divine purpose.

> For if we explain omnipotence as the attribute in virtue of which all finite things are through God as they are so we have certainly posited the whole divine act [*That*] but without motive [*Motiv*], hence as an absolutely undetermined action [*Handlung schlechthin unbestimmt*], and it can only be improperly, under the influence of quantitative standards of measurement, that we name God omnipotence.[155]

The notion that God is *merely*, "almighty and eternal is nothing more than the shadow of faith which even devils may have."[156] Instead, God is only truly known as omnipotent *love*.[157] And as Schleiermacher explains, "The divine wisdom is the principle which orders and determines the world for the divine self-imparting which is evinced in redemption."[158] It is the self-impartation of God through Christ that renders divine power determinate by assigning a definite purpose to God's activity.

The world is ordered as means to this end. "For our Christian self-consciousness," Schleiermacher says, "all other things are existent only in reference to the

155 *GL* §167.2; *The Christian Faith*, 730–731. Translation revised.
156 *GL* §167.2; *The Christian Faith*, 731; See: James 2:19.
157 *GL* §§166, 167; *The Christian Faith*, 727–32.
158 *GL* §168; *The Christian Faith*, 732.

efficacy of redemption [...]."[159] And, as with Leibniz, the determinate nature of the end renders the means determinate as well:

> Hence we can say, regarding two points formerly made out—both the essence of things in their relations to each other and the order of reciprocal interaction between them—that they exist through God as they exist with regard to the redeeming revelation of God in Christ, by which the spirit is developed to perfection. Everything in our world, namely, human nature in the first place and all other things in direct proportion to the closeness of their connection with it, would have been arranged [*eingerichtet*] differently, and the entire course of human and natural events, therefore, would have been different, if the union of the Divine Essence with human nature in the Person of Christ, and, as a result thereof, the union of the Divine Essence with the fellowship of believers through the Holy Spirit, had not been the divine decree [*Rathschluß*].[160]

Since the divine purpose is one purpose and not another, the world is disposed one way and not another. Since no means are indifferent to their end, and since the divine purpose is *determined* as this one thing (the impartation of the divine essence), the world as ordered to that purpose is accordingly determined.

Schleiermacher's position supports this account with a tacit subscription to Leibniz's notion of *conceptual containment*[161] (a point which will be explored in more detail in later chapters, particularly Chapter 5). That is, his position implies that no concept can be completely comprehended apart from the totality of its predicates. For example, the concept "Socrates" is not completely comprehended unless the predicate "who was executed by the Athenians," and all other predicates which are true of Socrates are understood as contained within the concept "Socrates." To completely understand what the concept "Socrates" means, it is

---

159 *GL* §164.1; *The Christian Faith*, 723. Translation revised.
160 *GL* §164.2; *The Christian Faith*, 724. Translation revised.
161 Like many of Leibniz's doctrines, conceptual containment can be construed more narrowly or more broadly. At its most specific, it is a doctrine of the relation of predicates to their subjects. I do not claim that Schleiermacher holds to this narrower view. However, the *basis* of conceptual containment in Leibniz's thought seems to be that all predicates of a subject are completely contained in the concepts of said subject because the subject *actually has* a relation to everything, and that anything true of a subject must be so because there is something *in* the subject that makes its concept true of it. This amounts to something much broader than a way of relating subjects and their predicates. It is a doctrine of the exhaustive mutual determination of all things. And it is this broader notion of conceptual containment, emphasizing the basis of the subject-predicate doctrine on the part of things, and their relations, that I principally have in mind when I claim that Schleiermacher subscribes to the doctrine. As we will see, Schleiermacher ascribes to the exhaustive mutual determination of all things in no uncertain terms. For a discussion of conceptual containment along these lines see Daniel Garber, *Leibniz: Body, Substance, Monad*, (Oxford: Oxford University Press, 2009), 182–89, especially 185–86.

necessary to understand all that Socrates ever was or did. The consequence of such a doctrine is to make the so-called accidental relations of things to be indispensable, to be relations that make the thing in question what it is and no other. And since any one thing has *some* relation to everything else, there is no complete comprehension of any one thing without exhaustive comprehension of that thing's relation to all other things. In the case of divine purpose and the constitution of the world, that means there is no complete comprehension of "the world" without the predicate "which is ordained to the redemption accomplished by Jesus of Nazareth."

In addition, conceptual containment will become important in later chapters in three ways. First, Schleiermacher uses this very doctrine of Leibniz's to argue against his theory of possible worlds. This is an important difference which will be addressed in Chapter 5. Second, conceptual containment is critical to answering the theodicy worries that naturally follow from exhaustive divine determination (see Chapter 6). And third, conceptual containment plays a key role in Schleiermacher's rendering of the eternal covenant as a blessing (also see Chapter 6). In each case, Schleiermacher, like Leibniz, relies on conceptual containment, as the corollary to the divine determination of all things, to support his account.

However, Schleiermacher finds himself, with Leibniz, facing the possibility that God might act for a determinate end with determinate means, but within or upon an already existing world. That is, in Leibniz's phrasing, God might determine ends and means but *not* "all their circumstances."[162] This possibility Schleiermacher likewise rejects:

> The word 'govern' [*regieren*] means to set in motion, and direct, forces which are already otherwise present; hence the term may easily mislead us into thinking even here of a divine direction of earthly forces conceived of as already existing, as also into separating the government of the world from creation in such a way that it appears like something later interjected [*hintennach zwischeneingekommenes*], and as if from the creation onward everything could have gone differently than in point of fact it has gone. The Christian faith that all things were created for the Redeemer implies on the contrary that by creation all things (whether as prepared for or as overruled) were disposed with a view to the revelation of God in the flesh, and so as to secure the completest possible impartation thereof to the whole of human nature, and thus to form the Kingdom of God. Similarly, the world of nature is not to be considered as going its own way on the strength of the divine preservation, the divine government only exerting influence through special isolated acts, so as to bring it into harmony with the kingdom of grace. To us, rather, the two things are absolutely one, and we have the certainty that from the beginning the whole disposition of nature would

162 Leibniz's Fifth Paper, in Alexander, *The Leibniz-Clarke Correspondence*, 78.

> have been different if the human race had not been determined [*bestimmt*] after sin to redemption through Christ.[163]

Instead, the totality of things were created and are sustained for a definite purpose, a purpose which is not later imposed, but immanent in the order and connection of things from their beginning.

Schleiermacher's case, again like Leibniz's, relies on the affirmation of perfection in God, and the corresponding removal of limits. The basis for his argument is found in his discussion of the divine attributes in general.[164] It begins with an ordinary definition of wisdom. By 'wisdom' [*Weisheit*], Schleiermacher says, "is understood the right ordering of plans [*Entwerfung*] to purposes [*Zwekkbegriffe*]—these considered in their manifold characteristics and in the totality of their relations to one another."[165] Anything done wisely includes the consideration of the *total relation* of means to ends. In the Christian consciousness of God, God's purpose is revealed in the redemption accomplished by Christ as the self-impartation of the divine nature, that is, as love. And it is for God, as for human agents, a perfection to do things wisely. In fact, as the removal of limits and affirmation of perfection suggests, God is *perfectly* wise. "Hence," says Schleiermacher, "since the divine government manifests itself in the self-consistent ordering of the whole sphere of redemption, alongside of the divine love we rightly place wisdom as the art [*Kunst*] (so to speak) of realizing the divine love perfectly."[166] And so:

> Without ascribing any limitation to God, therefore, we may assert that the divine wisdom is not fit to determine any other disposition of things, or any other ordering of their course, than the one wherein the divine love is *most perfectly realized*; and just as little is the divine love capable of leading to self-impartations other than those in which it finds perfect satisfaction, and so does not appear as absolute wisdom.[167]

Without recourse to the language of 'best', Schleiermacher argues in same vein as Leibniz that the perfection of the world is necessarily implied by the perfection of God as supremely wise.

Nevertheless, Schleiermacher denies that the notion of means *in an ordinary sense* can be employed in talk of divine action. Yet his denial takes a very specific form and is itself a negation only insofar as it is a further denial of limits and

163 *GL* §164.1; *The Christian Faith*, 723. Translation revised.
164 *GL* §50; *The Christian Faith*, 197–99.
165 *GL* §165.1 *The Christian Faith*, 727.
166 *GL* §165.1; *The Christian Faith*, 727.
167 *GL* §165.2; *The Christian Faith*, 727. Emphasis added. Translation revised.

affirmation of perfection. As Schleiermacher warns, "Next, we must be on our guard lest we again falsify the conception [*Begriff*] we have reached, by introducing the contrast of ends and means."[168] The word "contrast" is key, and this topic will be taken up again in greater detail in Chapter 6. For our present purposes, however, Schleiermacher offers two reasons why such a contrast in divine action is inadmissible. In the first place, "Every human artwork is the more perfect, the more it so conforms to this concept [*Begriff*], that, within the same, ends and means do not fall into this contrast, and the means only lie outside of it, but all within it behave as part to whole."[169] And since the perfection of the work of a human agent consists in the unity of ends and means, "How then could it be that divine wisdom not exclude this contrast even more completely?"[170] While it might not be obvious without additional reasons that a perfect work of art excludes this contrast, certainly if the perfection of creaturely art *did* consist in unity of ends and means, a perfect work of divine art must do so all the more.

Schleiermacher further offers a reason why perfection consists in the dissolution of the contrast of ends and means, a reason already hinted at when he said that "means remain external to it [the work of art]."[171] He explains that "means are only ever employed, where the agent must recourse to something that does not originate from himself."[172] That is, whereas an imperfect agent acts upon already existent material, a perfect agent is not only the shaper of the material but its creator. In this sense, God is obviously the one and only absolutely perfect agent. And if means are, by definition, not originated by the agent who uses them, then the use of means by God must be denied. Otherwise God would merely work upon an already existing, and so uncreated, something. Schleiermacher's point is plain enough. The end to which things have been ordered, and their creation, are not to be divided. If ends were imposed upon already existent matter, the divine art would be limited as above, and, thus, God's perfection denied. And, of course, God could not create means apart from ends, since there would be no sufficient reason to do so. Such an action would be utterly unintelligible. Schleiermacher's insistence on the indivisibility of creation and providence, not only secures for him a defense against a charge like Clarke's, that his God is a ruler in name only, but also supplies one of the critical premises necessary to generate the same dilemma that Leibniz sought: that

168 *GL* §168.1; *The Christian Faith*, 733.
169 *GL* §168.1; *The Christian Faith*, 733.
170 *GL* §168.1; *The Christian Faith*, 733. Translation revised.
171 *GL* §168.1; *The Christian Faith*, 733.
172 *GL* §168.1; *The Christian Faith*, 734. Translation revised.

God's supplementation or emendation of the natural order was always a *self*-supplementation. If the natural order is already ordained to a purpose in its inception, then supplementation or emendation necessarily implies either the definite insufficiency of that order to its ends, or that the natural order, though potentially sufficient, might still fail to fulfill its purpose. Schleiermacher's insistence on the unity of ends and means for God eliminates the first possibility.

The unity of ends and means eliminates the possibility of the insufficiency of the natural order by tethering the reason for the creation of the world to the end it is ordained to achieve. Thought through chronologically, from origin to purpose, the reason *why* a thing is made is to be or do something. But, the tethering of origin to *telos* also requires that the relation between the two be teleologically ordered when thought through in the other direction. That is, granted the reason for a thing is its end, how do its antecedent states contribute to that end? If the world's existence is, in fact, *for* its end, in order for its *origin* to be intelligible, then its existence must play a part in that end. Otherwise sufficient reason will not have been given for the world's existence. If the world contributes nothing to its end, it may as well not exist. Formal and final causes are coordinate. Formal causes, the kind of thing that a thing is, are defined in relation to final causes, the end for which a thing exists. What both Leibniz and Schleiermacher have done is argued that *the world as a whole*, if ordained to some end, must be of a certain nature, must be informed in a certain way, and so must be arranged one way and no other. That natural order, that formal cause, must be *precisely* the way it is because it is *perfectly* ordered to some final cause, some determinate end.

Formal causes are developmentally ordered to final causes. That is, the kind of thing a thing is *brings about* what it is for. Whatever contribution formal causes make to a thing's end, that contribution must be assigned as a genuine, rather than merely apparent, contribution. That means that antecedent states must causally influence subsequent states. And *that* requires that antecedent states must include *inherent* forces and powers (a theme we take up in the following chapter).

In the case of the world, there is no created thing, outside, to help or hinder it. If a finite cause could help or hinder, it would contribute to the causal nexus and so be part of the world after all. And so the world must, as a whole, include the totality of powers and forces which contribute to (and partially constitute) the fulfillment of its end. And, at that point, the only remaining question is: does or does not the world include within itself sufficient powers and forces to accomplish the end for which it was made? One could, of course, at this point simply claim that the world does not, or does so only imperfectly. But, such a qualification simply throws the question of the teleological ordering of the world back to square one. Since the *creation* of the world is only intelligible in light of its ordination to an end, if the world is insufficient means to that end, why create the world at all?

Whatever the world is meant to attain, if it is to be intelligible, *it*, the world itself, must so attain. And if the world is meant to attain, it should attain well. Otherwise its maker lacked power or wisdom or both. Schleiermacher, like Leibniz before him, regards the end for which the world was made as the condition of the intelligibility of God's action in creating the world. In consequence, its means-to-ends ordering is indispensable. And, since means are ordered to ends, they may be ordered better or worse in measure of how well they accomplish what they are for. But for God, as the perfect agent, nothing less than absolute perfection is admitted in regard to either ends or means. Hence, the world must be regarded as *perfectly* teleologically ordered as means to redemption. And, as above, whatever means contribute to ends must be inherent. In this sense, the world was made *inherently perfect:*

> As to what in the sphere of experience we call perfection or imperfection, the former is simply that which *by means of the original perfection* has already come to pass, the latter that which has not yet come to pass *by the same means*; both taken together, however, are the perfection which is coming to pass. Hence we can say that for each given moment the original perfection is in that which underlies it as pure finite causality; but the definitive perfection is in the totality of all the effects thereof, the development being thought of as included in the moment.[173]

That is, the original perfection of the world, and humanity as part of it, [174] consists in its being perfectly fit means to the divine ends (as above). That fitness lies inherent in things as the powers and forces which comprise their nature and which develop partially from within things, partially through the powers and forces resident in other finite beings, yet always from the world's own organic potential endowed along with its existence, "as the source of the whole temporal development."[175]

The inherent sufficiency of the world as means to its ends secures its perfection. And yet, that perfection might still be seen as a fragile one. It might be the sort of perfection which could be altered, disrupted, or diminished through the actions of free finite agents, particularly through sin. That is, the world could be *potentially* perfect, but not actually so, if its original perfection did not al-

173 *GL* §57.2; *The Christian Faith*, 235. Emphasis added.
174 *GL* §57; *The Christian Faith*, 233–36.
175 *GL* §57.2; *The Christian Faith*, 235.

ready include freedom and sin as a part and parcel with that perfection. If this hypothetical were granted, Schleiermacher's case, like Leibniz's, would be severely weakened. In that case, isolated divine acts would easily cohere with divine wisdom since God's interposition would *not* be a self-supplementation, but would rather correct activities that fall outside of the world order, activities which are themselves failures of perfection on the part of other agents. Since it is not an imperfection to rectify failure, God's special action in such cases would come as no surprise.

However, Schleiermacher admits of no instances where *any activity whatsoever* falls outside the divine causality. As he says:

> For, indeed, the capacity to change what has been ordained is only a merit in the ordainer when there is a need to change [*Aendernmüssen*], which again can only be the result of some imperfection in him or his work. If such an intervention [*Eingreifen*] be postulated as one of the privileges of the Supreme Being, it would first have to be assumed that there is something not ordained by Him which could offer Him resistance [*Widerstand*] and thus intervene [*eingreifen*] on him and his work; and such an idea would entirely destroy our fundamental feeling. [...] It follows from this that the most perfect explanation [*Darlegung*] of omnipotence would be a view of the world which made no use of such an idea.[176]

Just as Leibniz required divine power to argue from divine wisdom, Schleiermacher upholds both as well. No activity falls outside the divine activity, including the activity of free causes. "Whether or not," Schleiermacher says, "that which arouses our self-consciousness and consequently influences us, is to be traced back to any part of the so-called nature-mechanism [*Naturmechanismus*] or to the activity of free causes [*freier Ursachen*]—the one is as completely [*vollkommen*] ordained by God as the other."[177] Since free causes fall within the divine activity, there is no way they can fail to play their part in the divine purpose. They are themselves already included in the means and circumstances which contribute to the divine end. Consequently, no attempt to render the perfection of the world merely potential can succeed by appeal to free causes once their absolute dependence on God is granted. No, whatever the freedom of a free cause amounts to, it

> is always so conditioned that it only becomes what it is because it belongs to the very same universal system [*allegemeinen Zusammenhang*] which is the essential indivisible subject of the feeling of absolute dependence; and this would lose its significance in the whole prov-

176 *GL* §47.1; *The Christian Faith*, 179. Translation revised.
177 *GL* §49; *The Christian Faith*, 189.

> ince of history if we should think of free causes [*freier Ursachen*] as excluded from this system [*Zusammenhange*].[178]

If everything, including free causes, falls within the nature system, then appeal to forces outside the divine causality cannot be made to justify supplementation or emendation of that world.

Conclusion

Leibniz's argument given in his correspondence with Clarke, and Schleiermacher's position in *The Christian Faith* are, in the relevant respects, nearly identical. Both are arguments against discrete supernatural acts on the basis of divine wisdom. Both argue that since an action is unintelligible apart from an end, and since ends are always one thing and not another, God's action is unintelligible apart from a *particular* end. Both argue that means are also always determinate and so never indifferent to an end. Both argue that since God is a perfect agent, the perfection of means to ends follows necessarily. And further, both argue that as a perfect agent God not only orders things but also creates them, and so all circumstances and requisites are rendered determinate and perfectly ordered along with the means to God's determinate end. Both then argue that the inclusion of ends, means, and circumstances renders the world inherently perfect, since from its origin it is all-sufficient to its ends. And last, both argue that the wisdom of God in no way detracts from the power of God, and because all that happens falls within the sphere of divine omnipotence, there is no question of the world falling short of its purpose through the actions of free agents. For both, the result is a world which develops *solely* through the powers and forces inherent within it. Any attempt to append special divine acts to such an order impugns the divine wisdom, the divine power, or both.

In addition, Schleiermacher and Leibniz put the argument for the perfection of the world to the same uses. On the one hand, their argument leverages theological commitments in support of a general view of divine action and the natural world. It is the sort of world where natural explanations can always be offered *in principle*. And, since their general commitment to the naturalness of explanations excludes explanations of natural phenomena that appeal to supernatural causes, Schleiermacher and Leibniz hold a view of the world which denies authority to certain sorts of explanations, like Newton's, while it lends cre-

178 *GL* §49.1; *The Christian Faith*, 189.

dence to other sorts of explanations. It is, in other words, a view of the world in general which can be used to arbitrate between particular scientific theories. And in the case of Newton's natural philosophy, that is exactly the use to which it was put. As we saw in Chapter 2, Schleiermacher also subscribed to specific scientific theories which follow directly from this commitment. Again, just like the natural science of the day, higher and lower principles, principles and content, are coordinate.

On the other hand, Leibniz's and Schleiermacher's argument is a theological argument with decidedly theological implications. Theirs is an argument for one account of God over and against others. Recall that Leibniz's original objection arose not simply over Newton's natural philosophy but over *the kind of God it implied*. In contrast, both Leibniz and Schleiermacher firmly place divine power in service of divine wisdom. Neither exempts God from the requirements of the principle of sufficient reason under which reference must always be made to the good in determining action. Both reject claims like Clarke's that, "this sufficient reason is oft-times no other, than the mere will of God."[179] And all of this is done in the attempt to do adequate justice to the perfection of God.

Here on the doctrine of God, however, Schleiermacher and Leibniz part company. For while Leibniz pairs his use of the principle of sufficient reason with hypothetical counterfactuals in God, culminating in his doctrine of possible worlds, Schleiermacher rejects all use of hypothetical counterfactuals in God whatsoever. In consequence, he rejects Leibniz's distinction between hypothetical and absolute necessity (a subject we take up again in detail in Chapter 5). In lieu of Leibniz's conception, he adopts Spinoza's view that God does all God does by natural necessity. And yet, in spite of this divergence from Leibniz, Schleiermacher never gives up a teleologically ordered world and the appeal to divine wisdom. And as we will see, Schleiermacher's adoption of Spinoza does not undermine the basis of the eternal covenant, but instead provides him with powerful new tools to broaden and deepen his case. Spinoza not only gives his own distinct argument against discrete supernatural acts, but Schleiermacher uses Spinoza's view of all divine activity as free *and necessary* to secure the true benefits of the covenant. But before we turn to Spinoza, we will return to Schleiermacher's notion of a world. If the nature system is perfect, and if its perfection excludes miracles, both Schleiermacher's notion of the nature system and his definition of miracle will need to be explained.

---

179 Clarke's Second Reply, in Alexander, *The Leibniz-Clarke Correspondence*, 20.

Chapter 4 The World and Miracles: Schleiermacher on the Nature System

"First then, in my judgment, we must make a distinction and ask, What is that which always is and has no becoming; and what is that which is always becoming and never is?"

"All these are to be reckoned among the second and cooperative causes which God, carrying into execution the idea of the best as far as possible, uses as his ministers."

- Plato, *Timaeus*

The natural world as a unitary order is crucial to Schleiermacher's account. But how is that order known? Once again, for Schleiermacher, Christological determination of the world is necessary for the world to be known completely. Critically, however, it is not sufficient:

> If we look at the way in which we become aware of the two attributes respectively, it turns out that we have the sense of divine love directly in the consciousness of redemption, and as this is the basis on which all the rest of our God-consciousness is built up, it of course represents to us the essence of God. But the divine wisdom does not enter consciousness thus directly, but only as we extend our self-consciousness (as personal, but even more as consciousness of our kind) to cover the relation of all moments to each other.[180]

The *wise* ordering of the world is something we only come to by considering the world in its ordered structure.

In fact, nearly all of Schleiermacher's account relies on the notion of a *world*, either to complete his arguments, especially his arguments against divine intervention, or to specify what those arguments imply. In this chapter we examine in more detail what a world is, what the notion entails, and how Schleiermacher thinks that consciousness of the world, as a world, is connected to, but distinct from, the Christian consciousness of redemption. In this process, Schleiermacher also gives an additional argument against divine intervention, in passing, through the very notion of a world. Without a detailed understanding of these key terms, it is impossible to appreciate how rich Schleiermacher's account is, and it is also impossible to grasp the full force of the other arguments Schleiermacher gives in Chapters 3, 5, and 6.

180 *GL* §167.2; *The Christian Faith*, 732. Translation revised.

DOI 10.1515/9783110542301-004

In the process of displaying the strengths of his account, a number of objections will naturally suggest themselves. Besides making Schleiermacher's case clearer, this chapter aims to see off a number of the most popular objections to Schleiermacher's account. First, I show that Schleiermacher's notion of a system of nature is just as conceptually rigorous as his doctrine of God and, therefore, cannot be easily dismissed as a mere assumption, or as question-begging. Second, I demonstrate that the alternatives to Schleiermacher's account themselves entail unexamined burdens which his critics have yet to reckon with. And, finally, I argue that any accusation of Schleiermacher confusing or conflating God with the world reflects a deep misunderstanding on the part of his critics. It is, I argue, the critics of Schleiermacher's account of the nature system who are, ironically, most at risk of making God a mere thing among things, or of collapsing God and the world.

The System of Nature

The central role of the nature system [*Naturzusammenhang*] in Schleiermacher's account is as well-known as it is controversial. The notion is widely recognized as doing a great deal of work for Schleiermacher.[181] And yet it is not always clear *why* exactly Schleiermacher thinks there is a system of nature, and *exactly* what such a system is. Many analyses of *The Christian Faith* take the nature system as a methodological starting point in Schleiermacher's account.[182] Others take it as an unsupported ontological assumption.[183] The nature system is neither.

That the world is a *world* is not an immediate deliverance of the God-consciousness.[184] Instead, "To be conscious of oneself as part of the world is the same thing as to find oneself placed in a universal nature system."[185] Distinct from (but related to) the consciousness of redemption specifically, and God generally, it is a world-consciousness, a consciousness of being within and a part of a system of nature. This is a result of "reciprocal interaction": the experience of being relatively independent and relatively dependent in relation to other

---

181 See Dole, *Schleiermacher on Religion and the Natural Order*; Gerrish, "Nature and the Theater of Redemption"; Sherman, *The Shift to Modernity* (New York: T&T Clark, 2005).

182 See Lamm, *The Living God*, 127–136

183 William A. Dembski, "Schleiermacher's Metaphysical Critique of Miracles," *Scottish Journal of Theology* 49, 4 (1996): 443–465.

184 *GL* §§32, 34; *The Christian Faith*, 132–33, 137–40.

185 *GL* §34.1; *The Christian Faith*, 138.

things.[186] Since everything stands in some relation to everything else, the parts of this whole comprise a unity, albeit a diverse and manifold unity.[187] "This oneness with the whole in each several part is essentially twofold," Schleiermacher says. "A feeling of dependence, indeed, so far as the other parts act spontaneously upon it, but also a feeling of freedom in so far as it likewise reacts spontaneously on the other parts. The one is not to be separated from the other."[188] That is, the world is a unity because its diverse parts are all relatively dependent on one another and relatively free in relation to one another.

Everything finite stands in some relation to everything else finite, more or less affected, more or less affecting. As Schleiermacher explains:

> Let us now think of the feeling of dependence and the feeling of freedom as *one*, in the sense that not only the subject [i.e. we who feel ourselves free and dependent] but the corresponding Other [i.e. that upon which we are free and dependent] is the same for both. Then the total self-consciousness made up of both together is one of *Reciprocity* between the subject and the corresponding Other. Now let us suppose the totality of all moments of feeling, of both kinds, as one whole: then the corresponding Other is also to be supposed as a totality or as one, and then that term 'reciprocity' is the right one for our self-consciousness in general, inasmuch as it expressed our connexion which either appeals to our receptivity or is subjected to our activity. And this is true not only when we particularize this Other and ascribe to each of its elements a different degree of relation to the twofold consciousness within us, but also when we think of the total 'outside' as one, and moreover (since it contains other receptivities and activities to which we have a relation) as one together with ourselves, that is, as a *World*.[189]

That is, if we are both free in regard to, and dependent upon, other things, then our relation to other things is one of reciprocal interaction. And if we do not artificially restrict the scope of the 'other' in relation to which we are both relatively free and relatively dependent, then the 'other' we stand in relation to is not only this or that particular thing, but the totality of things that share this mutual reciprocity:

> [T]owards all the forces of nature—even, we may say towards the heavenly bodies—we ourselves do, in the same sense in which they influence us, exercise a counter-influence, however minute. So that our whole self-consciousness in relation to the World or its individual parts remains enclosed within these limits.[190]

186 *GL* §32.2; *The Christian Faith*, 132.
187 *GL* §32.1; *The Christian Faith*, 131–32.
188 *GL* §32.2; *The Christian Faith*, 132.
189 *GL* §4.2; *The Christian Faith*, 14–15.
190 *GL* §4.2; *The Christian Faith*, 15.

When our consciousness is extended to embrace more and more, we find that there is no limit to the nature system short of the whole, but that it embraces all finite things.[191] *All* things in the world stand in a mutually conditioned and mutually conditioning relation.

However, the world is not *all there is*, but all *finite* things. "For we recognize in our self-consciousness an awareness of the world," Schleiermacher explains, "but it is different from the awareness of God in the same self-consciousness."[192] As above, we already know that the consciousness of God and the consciousness of the world are distinct. Is Schleiermacher simply repeating that they have different sources? No, they certainly do, but the consciousness of God and the consciousness of the world are also distinct in content. In each case we are conscious of categorically different things. On the one hand, the world is a manifold, a *relative* unity, an ordered diversity. And we are conscious of it precisely as diverse. God, on the other hand, is an "absolute undivided unity."[193] "[T]his 'Whence' [i.e. God]," Schleiermacher says, "is not the world, in the sense of the totality of temporal existence, and still less is it any single part of the world."[194]

This effort to distinguish God and the world resists the reduction of God to the world along the lines of what today is called 'metaphysical naturalism,' the notion that the world as an aggregate of finite things is all there is. "To be one with the world in self-consciousness," Schleiermacher explains,

> is nothing else than being conscious that we are a living part of this whole; and this cannot possibly be a consciousness of absolute dependence; the more so that all living parts stand in reciprocal interaction with each other. This oneness with the whole in each several part is essentially twofold: a feeling of dependence, indeed, so far as the other parts act spontaneously upon it, but also a feeling of freedom in so far as it likewise reacts spontaneously on the other parts. The one is not to be separated from the other.[195]

It is this very absoluteness of dependence which implies the aseity of God, and so funds the first set of divine attributes.[196] It also provides Schleiermacher with

---

191 "This system, however, is not posited as having limits; hence it contains within itself all finite being, only in undeveloped form." *GL* §34.1; *The Christian Faith*, 138.
192 *GL* §32.2; *The Christian Faith*, 132.
193 *GL* §32.2; *The Christian Faith*, 132.
194 *GL* §4.4; *The Christian Faith*, 16.
195 *GL* §32.2; *The Christian Faith*, 132.
196 *GL* §§50–56; *The Christian Faith*, 194–232. For example: "[A]ll the divine attributes to be dealt with in Christian Dogmatics must somehow go back to the divine causality, since they are only meant to explain the feeling of absolute dependence." *GL* §50.3; *The Christian Faith*, 198.

an infinite creator-creature distinction without need for a finite past or a finite future. There is an infinite distinction between the absolutely *a se* and the absolutely dependent, between a being whose essence involves existence and those beings whose existence is not their own. However, while Schleiermacher's distinction between God and the world does these things, it also does much more.

The world is distinguished from God by the mutual affectability of its causes. Schleiermacher explains that:

> The feeling of absolute dependence, accordingly, is not to be explained as an awareness of the world's existence, but only as an awareness of the existence of God, as the absolute undivided unity. For neither is there in relation to God an immediate feeling of freedom, nor can the feeling of dependence in relation to Him be such that a feeling of freedom can be its counterpart.[197]

Our consciousness of God is not to be confused with our consciousness of the world because God is distinct from anything in the world, and the world as a whole, in that God is uniquely that upon which we are absolutely dependent.

We are aware of this difference because, although we stand in some relation to all finite things (and so the world as a whole) and to God, only in the case of God do we stand in a relation of *absolute* dependence, and only in the case of finite things do we stand in a relation of relative freedom.[198]

> Accordingly our self-consciousness, as a consciousness of our existence in the world or of our co-existence with the world, is a series in which the feeling of freedom and the feeling of dependence are divided. But neither an absolute feeling of dependence, i. e. without any feeling of freedom in relation to the co-determinant, nor an absolute feeling of freedom, i. e. without any feeling of dependence in relation to the co-determinant, is to be found in this whole realm.[199]

And, as above, we know we stand in a relation of relative freedom to finite things because we can act *upon* them. As Schleiermacher says:

> For we have a feeling of freedom (though, indeed, a limited one) in relation to the world, since we are complementary parts of it, and also since we are continuously exercising an influence on its individual parts; and, moreover, there is the possibility of our exercising our influence on all its parts; and while this does permit a limited feeling of dependence [on the world], its excludes the absolute feeling.[200]

---

197 *GL* §32.2; *The Christian Faith*, 132.
198 *GL* §32.2; *The Christian Faith*, 132–33.
199 *GL* §4.2; *The Christian Faith*, 15.
200 *GL* §4.4; *The Christian Faith*, 16–17.

On the other hand, the feeling of absolute *freedom* is, likewise, excluded because everything also acts upon us. As Schleiermacher again explains:

> There can, accordingly, be for us no such thing as a feeling of absolute freedom. He who asserts that he has such a feeling is either deceiving himself or separating things which essentially belong together. For if the feeling of freedom expresses a forth-going activity, this activity must have an object which has been somehow given to us, and this could not have taken place without an influence of the object on our receptivity. Therefore in every such case there is involved a feeling of dependence which goes along with the feeling of freedom and thus limits it. The contrary could only be possible if the object altogether came into existence through our activity, which is never the case absolutely, but only relatively.[201]

In order for us to be absolutely free, our activity would have to be wholly unconditioned by any extrinsic object. But that is impossible: in order to be absolutely free we would have had to be the absolute creator of such an object, and only God brings things into existence absolutely.

The mutual reciprocity of all finite things is in contrast to God who is wholly independent, and upon whom we are absolutely dependent—who is wholly unaffected by the world and instead wholly conditions it. Finite things make one another to be what they are. They are dependent on each other at the same time that they are also relatively free in regard to one another. Finite things are contingent, not in the sense that God might not have made them (see Chapter 5), but in the sense that their reason for being, and for being as they are, is not their own. Both for their sheer existence, and for their existence as they are, they depend on all other things, and on God. God, in contrast, is uniquely necessary in both senses. God is solely determined to be and to act by the divine essence alone. God is wholly independent, that on which everything absolutely depends and which itself depends on nothing.[202] In consequence, the divine existence is not compound. It is not the result of itself plus anything else. God is wholly *a se*, and so entirely simple, eternal, immutable, and free since uniquely, entirely, and necessarily determined to act by the divine nature alone.

In consequence, it is impossible that there be a thing which is part of our world which is neither mutually conditioned by, nor mutually conditioning of, all other finite things. On the one hand, if a thing were not mutually conditioning then it could not be part of *our* world because even to learn of it would condition our knowledge. For it to even *be* it would have to be in some relation to other

---

201 *GL* §4.3; *The Christian Faith*, 15–16.

202 "But if the simple expression that 'everything depends upon God,' is further supplemented by the negative 'but He Himself upon nothing,' at once a fresh opening is given for a division into positive and negative attributes." *GL* §50; *The Christian Faith*, 199.

finite things; and if in some relation it is necessarily mutually conditioning—even if the relation is a supposedly passive relation such as in knowledge. As Schleiermacher says:

> Even if we were known to ourselves only as presentational activity—that is to say, as being centres for ideas—even so the self-consciousness is a centre for truth; and that implies a relation of being in the self-consciousness, corresponding to the relation of ideas in the objective consciousness.[203]

That is, ideas, no less than things, are included in the mutual affectability of the finite.

However a thing *is*, if it is part of the world, it conditions the world and the world conditions it. And conversely, if a thing does not condition the world and is not conditioned, it is not part of it.

On the other hand, we might imagine a thing which does condition *other* things but which is not itself conditioned. But to be truly unconditioned it must be wholly unconditioned. That is, it must not stand in a conditioned relation with anything else whatsoever. It must be absolutely self-conditioned. But that would make such a thing *a se*, self-subsistent, the reason for its own existence, and that is the definition of God. But a plurality of aseities, a plurality of Gods, violates the feeling of absolute dependence upon which Christian piety is based, not to mention running into the logical absurdity of a plurality of aseities (i.e. 'substances') as Spinoza points out.[204] Therefore, there can be one and only one absolutely unconditioned *Whence* of it all, while *everything* else is mutually conditioned and mutually conditioning.

Why is this important? The first of several reasons is that it makes plain that Schleiermacher's account of the world is not merely a description of the order of discovery but also a description of the order of being. In consequence, Schleiermacher's commitments are not limited by his starting point in consciousness. However one comes to subscribe to the aseity of God and the dependence of creatures, Schleiermacher's arguments hold *ex hypothesi*. That is, granted that God and the world are distinguished as Schleiermacher (very traditionally[205]) argues, then the world *must* consist in the *totality* of finite things precisely because a finite thing *by definition* stands in some mutually conditioning and mutually conditioned relation to some other finite thing, and ultimately to every finite

203 *GL* §34.1; *The Christian Faith*, 138.
204 Spinoza, *Ethics* IP14; See also Anselm of Canterbury, *Monologion*, Ch 4; Thomas Aquinas *Summa Theologiae*, Ia, q. 11, a. 3, resp.
205 See Anselm, *Monologion*, Ch 7. Aquinas, *Summa Theologiae*, Ia, q. 11, a. 3, resp.

thing. If it is absolutely unconditioned, it is God; if it in no way conditions finite things, it has no actuality, and so no being. It simply doesn't exist (see below). This alone entails that the world has some minimal order, that in some sense it is a *world*. Schleiermacher will add to this further, but even with only this first step in place the world is already minimally known to be not less than the ordered co-existence of the totality of finite things—ordered, precisely because those things are, by definition, mutually affecting and affectable.

Though starting in consciousness, Schleiermacher nevertheless makes definite ontological claims, but by proceeding as he does he also dismisses a potential rival account on his way to these claims: occasionalism. A metaphysical alternative to Schleiermacher's account, occasionalism is the doctrine that nothing but God is a genuine efficient cause, that no creature is the proximate cause of its own effects.[206] Though denying creatures the power of genuine causality and reserving it only for God, occasionalism still holds that everything is absolutely dependent on God. However, by denying creatures causal efficacy, occasionalism denies that there is, properly speaking, a *world* in the sense described above because created things are not ordered to one another as mutually conditioned and conditioning, but only as absolutely conditioned by God. This is attractive to some because without a world in Schleiermacher's stronger sense, there is no possibility of miracles as "interventions" in that world. Instead of theorizing a way for God to act within or upon a world, the whole problem is circumvented by denying a system of nature in the first place.[207] Predictably, Schleiermacher wants nothing to do with this account. Occasionalism faces many burdens, not all of which will detain us here. However, by proceeding as he does, Schleiermacher exhibits several of these burdens in passing.

We are conscious of being absolutely dependent on God, Schleiermacher claims, in positive proportion to our awareness of being relatively free and relatively dependent.[208] By definition, those things upon which we are not absolutely dependent are not God. This distinguishes creatures from God such that neither God nor the world are collapsed into the other. Absent *some* distinction between creatures and God, the God-world relation resolves into total acosmism as easily

206 For the history and development of this doctrine see Daniel Garber, "Descartes and Occasionalism" and Steven Nadler, "The Occasionalism of Louis de la Forge" in *Causation in Early Modern Philosophy* (University Park, PA: Pennsylvania State University Press, 1993). For an account in its developed form see Robert Merrihew Adams, "Malebranche's Causal Concepts" in *The Divine Order, the Human Order, and the Order of Nature*, ed. Eric Watkins (Oxford: Oxford University Press, 2013).

207 See Dembski, "Schleiermacher's Metaphysical Critique of Miracles," 458–464, esp. 463.

208 *GL* §34; *The Christian Faith*, 137–38.

as it resolves into pantheism. And the distinction Schleiermacher offers, that of *relative* freedom and dependence, is not available to one who denies that created things are the proximate cause of their own effects. By maintaining that creatures are the genuine causes of their effects, yet no less dependent on God for it, Schleiermacher resists the dissolution of the world into God in a way that occasionalists cannot.[209]

Further, our status as agents is simply a fact of consciousness for Schleiermacher. It is through our ability to act, and recognize ourselves as centers of activity, that we become conscious of other things over and against ourselves, and eventually come to the recognition of the mutual relatedness of the whole.[210] It is through our experience of relative freedom and relative dependence, our experience as finite agents, that we become aware of the dependence of all finite being on God. It is not even possible, according to Schleiermacher, to think of ourselves as not acting since thinking itself is an action. Even if human being itself was nothing but a center for thought, we should still find ourselves in a relation of mutual affectability in regard to other finite things.[211] It is only in the experience of the world as a "divided and disjointed unity,"[212] implied by our actions and their limits, that we become aware that no particular thing in the universe, nor the universe as a totality, is that upon which we are absolutely dependent.

It is precisely our experience of finite things *as affectable* by other finite things, and God, in contrast, as wholly unaffectable, that we not only distinguish ourselves from other things, but also distinguish the world from God. There is an exact parallel between our *consciousness* of being agents and our existence as agents *in fact*, and between our consciousness of dependence on God and our absolute dependence on God *in fact*. To deny what is surely *prima facie* more plausible—that we are at least sometimes the genuine causes of our own actions—is to introduce an element of illusion which undermines not only the awareness of our own human agency, but also our consciousness of dependence on God. In that case, consciousness in general, and so our consciousness of God in particular, would become suspect and unreliable.

Nor are Schleiermacher's reasons for holding that finite things can be genuine efficient causes new. In fact, in the heart of debates over occasionalism in his own time, Leibniz excoriated his opponents for the same faults:

---

209 This is also what Rutherford calls Leibniz' first argument against occasionalism: that to deny the genuine activity of created things is to deny their substantial being. See Donald Rutherford, "Natures, Laws, and Miracles" in *Causation in Early Modern Philosophy.*

210 *GL* §46.2; *The Christian Faith*, 173–74.

211 *GL* §34.1; *The Christian Faith*, 138.

212 *GL* §32.2; *The Christian Faith*, 132.

> For who would call into doubt that the mind thinks and wills, that we elicit in ourselves many thoughts and volitions, and that there is a spontaneity that belongs to us? If this were called into doubt, then not only would human liberty be denied and the cause of evil things be thrust into God, but it would also fly in the face of the testimony of our innermost experience and consciousness, testimony by which we ourselves sense that the things my opponents have transferred to God [i.e. efficient causes], without even a pretense of reason, are ours.[213]

Moreover, Leibniz did not only cite the testimony of experience. He also pointed out that occasionalism's proponents unknowingly committed themselves to strange, and ultimately disastrous, notions of God. As Leibniz put it:

> [T]he very substance of things consists in a force for acting and being acted upon. From this it follows that persisting things cannot be produced if no force lasting through time can be imprinted on them by the divine power. Were that so, it would follow that no created substance, no soul would remain numerically the same, and thus, nothing would be conserved by God, and consequently everything would merely be certain vanishing or unstable modifications and phantasms, so to speak, of one permanent divine substance. Or, what comes to the same thing, God would be the very nature or substance of all things, the sort of doctrine of ill repute which a recent writer, subtle indeed, though profane, either introduced to the world or revived.[214]

The effect of the occasionalist denial of the intrinsic power of created things was, ironically, to dissolve the dependence on God that occasionalists hoped to preserve; not by suggesting that things did not depend on God, but by suggesting that, in the final analysis, there were no *things* at all. In place of a world that depended on God, there was nothing but a single substance and its phantasmal appearances. In short, occasionalists inadvertently became Spinozists.

Of course, it is not clear that Spinoza himself actually subscribed to the views that Leibniz attributed to the occasionlists. For instance, Spinoza thinks that *things* produce determinate effects without supplemental divine activity, and so actually agrees with Leibniz *against* the occasionalists on that count.

---

213 Leibniz, *On Nature Itself* (1698) in Roger Ariew and Daniel Garber, ed. and trans., *Philosophical Essays,* (Indianapolis: Hackett, 1989), 161. A conspicuous difference between Leibniz's position and Schleiermacher's might seem to lie in Leibniz's denial that God is the cause of evil. But this is not as different as may at first appear. For those who deny secondary causes, God is the *proximate* cause of evil, the particular cause of each and every evil occurrence. For Schleiermacher, however, God is the cause of no discreet events whatsoever within the world, but only of the world as a whole. And only insofar as the world as a whole includes evil is God the cause of evil for Schleiermacher. Therefore, both Schleiermacher and Leibniz agree that God is not the proximate cause of evil in this strong sense. See Chapter 6.

214 Leibniz, *On Nature Itself*, 159–160.

On the other hand, Spinoza denies that bodies are distinguished from one another by reason of substance.[215] Here the issue is what exactly Spinoza thinks a *substance* is. As we will see in passing in Chapter 5, Spinoza and Leibniz use the term "substance" in two very different ways. Nevertheless, even if Spinoza is not vulnerable to the accusations Leibniz levels against him, it does not follow that occasionalists are accordingly immune. They might hold *worse* views than Spinoza. For Schleiermacher, the issue is even clearer. He does not label either God or individual created things as "substance" or "substances," and so does not quarrel over the proper use of the word. For him, it is enough to note that created things are the genuine causes of their effects, that they are not mere phantasmal appearances, and yet are absolutely dependent on God.

In light of the reasons given by one of occasionalism's most vigorous opponents more than a century prior, Schleiermacher's ascription of order to the world is as far from naïve as revived objections from occasionalism are convincing. In fact, not only the content of his doctrines, but even their order of discovery seems deliberately imitative of Leibniz who, perhaps more than any other thinker, sent occasionalism into eclipse. In its place, Schleiermacher offers an account of finite beings as genuine causes, such that things have a determinate nature, a nature which explains their powers and actions. In so doing, he follows much of the theological tradition.

Schleiermacher's alternative is to distinguish between primary and secondary causes.[216] He claims that, "divine preservation, as the absolute dependence of all events and changes on God, and natural causation, as the complete determination of all events by the universal causal nexus, are one and the same thing simply from different points of view, the one being neither separated from the other nor limited by it."[217] That is, there are two causal layers that are ultimately identical, but those causal orders are nevertheless also to be distinguished. The ultimate identity of the two is implied by the absolute dependence of all on God. If anything were to fall outside that causality, it would violate that most basic premise. But, nevertheless, natural causality is not divine causality in the same way that God is not the world, in a sense of simple identity. Schleiermacher continues:

---

215 Spinoza, *Ethics*, IIL1.

216 The distinction between primary and secondary causes has a long theological history. It is, of course, most famously employed by Thomas Aquinas. See also Leibniz's, *Discourse on Metaphysics*, Section 8.

217 *GL* §46.2; *The Christian Faith*, 174.

> Moreover, in order not to confuse ourselves in this way, we ought to observe more carefully the difference between a universal and an individual cause [*allgemeinen und besonderen Ursache*]. For in the totality of finite being only a particular and partial causality is given to each individual, since each is dependent not on one other but on all others; the universal causality attaches only to that on which the totality of this partial causality is itself dependent.[218]

Universal and individual causes—Schleiermacher's terms for primary and secondary causes—are distinguished through the same contrast between the absolutely independent and absolutely dependent, which is the basis for Schleiermacher's distinction between God and world. Insofar as there are some things which are not God, that is, insofar as there are finite things, there are some things which stand in a mutually conditioned and mutually conditioning relation to other finite things. But in order for *those things*, individual causes, to be *genuinely* mutually conditioned and conditioning, they must possess "a particular and partial causality"[219] intrinsically. Individual causes must have the power to act. Otherwise, *they* would neither mutually condition nor be mutually conditioned. On Schleiermacher's account, individual causes must be particular things possessed of particular powers.

In turn, Schleiermacher holds, these powers must be explained by, and coordinate with, the beings of those things. "[P]reservation has as its object the being of things," Schleiermacher says, "and in this, so far as they are centres of power [*Ort für Kräfte*], the antithesis of self-activity and susceptibility is included […]."[220] In saying that divine preservation has as its object the *being* of things *in so far as they are centres of power*, Schleiermacher distinguishes his view from those who hold that divine cooperation has, as its special object, the *actions* of things. "[F]or some," Schleiermacher explains, "connect the expression 'preservation' only with matter and form [*Stoff und Form*], and 'co-operation' with powers and actions [*Kräfte und Handlungen*]; others again connect preservation with the existence and powers of things, and co-operation only with activity."[221] That is, some distinguish the existence of a thing, either as matter and form, or as power, from its actions and activity. But this distinction can suggest that the existence of things and activity of things do not equally fall under divine preservation. "This tendency," Schleiermacher says, "must be en-

218 *GL* §46.2; *The Christian Faith*, 174–175.
219 *GL* §46.2; *The Christian Faith*, 174–175.
220 *GL* §46 postscript; *The Christian Faith*, 178.
221 *GL* §46 postscript; *The Christian Faith*, 176.

tirely avoided and not merely covered up by indefiniteness."[222] He then continues:

> If, however, such a distinction ought not to be drawn, and if the powers of things are something as little separated from the divine sustaining activity as their being itself (the latter we only divide into matter and form by an abstraction which has no place here), then the difference between preservation and co-operation rests also on a similar abstraction. For *being posited for itself can only exist where there is also power, as power always exists only in activity* [*Denn ein für sich zu sezendes Sein ist doch nur da wo Kraft ist, so wie Kraft immer nur ist in der Thätigkeit*]; thus a preservation which did not include the placing of all the activities of any finite being in absolute dependence on God would be just as empty as creation without preservation.[223]

On Schleiermacher's account no supplemental divine activity is required for things to *act* once they *are*. In preserving the being of things, God preserves things precisely *as* centers of power, which is to say, as things that *can* and *do* act, as things that have an *innate* power to act. And since this activity is already included in the being of things, the powers and actions of things do not fall outside the divine preservation. Therefore, God's special cooperative power is not needed to bring about the effects of finite things, "so that deeds, as distinct from the preservation of powers, proceed from a divine activity,"[224] because God's preservation of things already includes preserving them *as* centers of power and action.

In addition to his identification of the being and powers of things, Schleiermacher also uses the notion of a thing's *nature* to explain the kinds of actions a thing performs. "[T]he individual can act only in accordance with the nature of his species," Schleiermacher says, "but never can act upon that nature."[225] And although Schleiermacher does not explicitly equate the nature of a thing with its being, it would be a natural extension of what Schleiermacher has claimed, not to mention traditional, to do just that and identify the *being* of a thing with its *nature*, and, thus, to make both the kind of being in question, and its powers, determinate. Put a little differently, if it is the being of a thing as a center of power which explains what that thing can and does do, the language of natures would merely lend further specificity to this account by adding that the being of a thing will be a particular being, disposed to act in a particular way. And in-

222 *GL* §46 postscript; *The Christian Faith*, 176.
223 *GL* §46 postscript; *The Christian Faith*, 176. Emphasis added.
224 *GL* §46 postscript; *The Christian Faith*, 177.
225 "[...] *da es ja immer nur mit der Natur seiner Gattung handeln kann, niemals aber auf dieselbe.*" *GL* §72.3; *The Christian Faith*, 296.

deed, Schleiermacher makes wide appeal to the natures of things in other contexts, most prominently, human nature, to explain powers and actions characteristic of a kind. It is the nature of a thing which explains the powers it intrinsically possesses. That all things are one thing, and not others, means that they do some things, and not others. They have determinate natures, specific forms, suited to particular activities. Below we will see how this enables Schleiermacher (again following Leibniz) to specify exactly what a miracle is, and to further distinguish between "relative" and "absolute" miracles. But, before we do, we return to the nature system in general, and this time consider the world as a whole, that is *itself* a specific kind of being, able, and indeed ordained, to produce effects coordinate with its nature.

Schleiermacher extends the notion that the being of a thing *includes* its powers and activities to the world as a whole. And, just like the natures of individual things, the world too has a nature, disposing it to particular ends. As we saw in the previous chapter, redemption in Christ renders divine action determinate by specifying the *telos* of the world as love. And not only is the world a necessary condition of this end, it is also a sufficient one. Not only is love what the world is *for*, but love is what *the world* is for. For the world to have been made with sufficient reason, it must be sufficient to the end for which it was made. One implication is that miracles don't happen because they needn't happen. The world is already perfect as means to end. But this is not the only implication. Another is that the world, since perfectly ordered to a determinate end, is of a nature, of a form, perfectly coordinate with that end. The reason for any thing's *being* is its reason for its *being in a particular way.* And this is not only true of individual things, but of the world as a whole. The world is doubly ordered: as finite thing to finite thing and as finite things to God. On both accounts the world is a *world*, a unitary order.[226] The nature system is itself a *thing*, arranged and ordered to an end. And divine wisdom perfectly coordinates the world as formal cause with redemption as final cause.

Additionally, on Schleiermacher's account, the world as a world, since wholly sufficient to its end, must possess intrinsic powers coordinate with its end, just like individual things. And, in fact, since the world is *perfectly* ordered to its end, the world as a world must intrinsically possess *all* the power necessary for the attainment of the end for which it was made. Again, this implies that the world itself has a form—which is to say that the world as a world is a natural *order* in the strongest sense. Just as finite things have natures which explain how they are the genuine causes of their own actions, actions which are charac-

226 See Aquinas, *Summa Theologiae*, Ia, q. 47, a. 3, resp.

teristic of the kind of thing they are, so the world, nature itself, also has a nature (in this other sense) which explains all that happens in it. Further, since the world *must* be perfectly sufficient to its end (or else violate the perfection of divine wisdom), unlike the finite things *in* the world, the world itself must be the sole and sufficient genuine source of its own effects. This not only follows from the perfection of the world, but its unity and uniqueness: there is no created thing that is not of the world, and there is but one world. In contrast, finite things are sometimes not the genuine cause of their effects. They are sometimes affected by other finite things. And they are *never* the sole cause of their effects because they act within and upon already given circumstances. The totality of finite things is mutually determining. The nature system, considered as a whole, is different. Because all finite things have a relation to one another and a relation to God, the world is a single order. It is, in the strictest sense, a *universe*. And, in consequence, the world cannot limit and be limited by the actions of other created natures, for there are no other created things to limit it. Nor do particular finite things affect the world. They are, rather, effects *of* the world, and the nature system is their cause.

However, the notion that the world is sufficient to produce its own effects might be taken to suggest that God is superfluous to the nature system, that somehow the nature system is self-subsistent. This could not be further from the truth. The nature system is no exception to the rule that everything is absolutely dependent on God for its existence. Nor is the nature system to be confused with the divine essence itself such that the nature system *is* that essence. For the world is not only an order of things to things, but also of all things to God. That is, as we saw in Chapter 3, Schleiermacher holds to the principle that no action is explicable apart from an end for which to act. Since, for Christians, the creation of the world is a divine act, it is only intelligible in light of a determinate end. Schleiermacher specifies this end as the divine love accomplished through the redemption of Christ. And so it is to the divine love (i.e. the divine essence) as final cause that the world is ordered as formal cause. And, in fact, the world is distinguished from God in direct proportion to the degree that it is ordered to God as something not itself. That is, insofar as all things are ordered to something *else* as their final end, they are not and cannot be their own end. There is some other good on which they depend, not only for their origin, but for their perfection. But, God does not depend on anything at all for God's perfection, but is perfection-itself.[227] Just as the divine aseity distinguishes

227 That Schleiermacher holds to the perfection of God is most clearly seen in his use of the *via negativa* where God's imperfection is, over and over, denied. For example, in the context of the

created things from God regarding their origins, so too the self-sufficiency of divine perfection distinguishes created things from God regarding their ends. The more the world as a nature system is ordered *to God* as something transcending itself, the more it is a world.

Nevertheless, the nature system, while absolutely dependent on God, is no less self-sufficient to produce its own effects. And this might continue to suggest to some that the world is somehow still *too* independent. But again, this is a false worry. The nature system as an order is no less dependent on God than anything else. Its reason for existence is in no way its own. Moreover, while the nature system is sufficient to produce its effects, it is not sufficient to produce its own nature. The world depends on God not just for its existence, but also for its essence, for being the kind of world it is—a world which takes the form it does as ordered to the divine end it was created for. In both regards, Schleiermacher's account of the world's dependence is clear when the principle that *things do not act upon their own natures, but in accordance with them,*[228] is reiterated (that is, when the principle of sufficient reason is couched in the language of natures). What happens in the world does so according to the nature of the world. And as regards both the kind of world it is, as well as its sheer existence, the world is absolutely dependent on God.

These distinctions help us explain a great deal in Schleiermacher's theology. They flesh out his concept of a nature system—both why he thinks that the world is such a system, and why he thinks it works in the way it does. As the totality of the mutually affectable, ordered in itself thing to thing and as a whole to God, the world is formed, arranged, and structured in a specific way for a specific purpose. The world is not first and foremost an orderless multitude of things only subsequently arranged. What happens in the world is a result of its order, not the other way around. Nor is the order of the world subsequently imposed on things that already exist in some other definite way. The world is no artifice. In-

---

attribute of omniscience Schleiermacher concludes: "Thus every distinction in content between wisdom and omniscience must presuppose an imperfection in God." Tellingly, that sentence *concludes* his argument, clearly implying that he regards the idea of imperfection of God as absurd. *GL* §55.1; *The Christian Faith*, 221. Further, if God is both perfect and *a se*, God must be perfection-itself since, as *a se*, God does not depend on anything else to be, or be a certain way. Whatever God is, God is essentially. And so if God is perfect, God is uniquely and essentially perfect, which is to say, perfection-itself.

228 Again, see Schleiermacher' use of this principle in the context of the Fall: "Still less is it possible to suppose that such an alteration of nature should have resulted from an act of the alleged individuals as such, since the individual can act only *in accordance* with the nature of his species, but never can act *upon* that nature." *GL* §72.3; *The Christian Faith*, 296. Emphasis added.

stead, its end is the reason for its order, and its order the reason for all that happens in it. Like a life, the world develops according to an intrinsic principle. Its *telos* is immanent in its nature, not in the sense that it is its own end, but in the sense that its reason for being, and for being the way it is, is its end. The nature system, the universe, is an *organism*, distinguished from all other life by its unity and uniqueness, and so its sufficiency to its end, and so its perfection. It is a *perfect living thing.*

A great deal of Schleiermacher's account of the distinction between primary and secondary causes owes itself to the broader theological tradition, including Leibniz. However, his doctrine of the world as a unitary thing also combines aspects of this account with features of Spinoza's account of nature. Like Leibniz, Schleiermacher describes the world as the totality, the aggregate, of finite things. The world is the whole of the mutually affectable. And this whole is teleologically ordered to God as ultimate final cause. But, like Spinoza, Schleiermacher thinks the world is not a *mere* aggregate, but rather an ordered whole. As Schleiermacher explains:

> For that feeling [of absolute dependence] is most complete when we identify ourselves in our self-consciousness with the whole world and feel ourselves in the same way as not less dependent. This identification can only succeed in so far as we unite everything that in appearance is scattered and isolated and, by means of this unifying association conceive of everything as one. For the most complete and universal interdependence of nature is posited as this 'All-One' of finite being, and if we also feel ourselves to be absolutely dependent, then there will be a complete coincidence of the two ideas—namely, the unqualified conviction that everything is grounded and established in the universality of the nature system, and the inner certainty of the absolute dependence of all finite being on God.[229]

Finite being itself forms an 'All-One' unity, a stronger notion of the 'world', more akin to Spinoza's view than Leibniz's.

In contrast to Leibniz, Spinoza gives an account of the world as a single *individual* (in Spinoza's technical sense). His account of the world as an individual begins with a notion of *bodies* applied to bodies of all kinds—from bugs, to people, to stars.[230] A body, composed either of "simples" or of other bodies, forms an individual when its parts "communicate their motions to each other in a certain fixed manner [*ratio*]."[231] There is no limit to the composition of larger and larger bodies in this way. And so, says Spinoza, "If we proceed in this way to infinity,

229 *GL* §46.2; *The Christian Faith*, 173.
230 Spinoza, *Ethics*, IIL2.
231 Spinoza, *Ethics*, IID.

we shall easily conceive that the whole of nature is one individual, whose parts, that is, all bodies, vary in infinite ways without any change of the whole individual."[232] The totality of all bodies forms an *individual* because they all communicate their motions to one another in a fixed *ratio*. To this view, Schleiermacher only need add the claim that the *ratio* of the world is perfectly ordered to an end, which, as we saw in Chapter 3, he does. Since the innate order of a thing to an end is its *form*, the universe as a whole, this infinite individual of all interconnected bodies, is united not only by a common *ratio*, but also a common *telos* in a common form.

Further, the unity of the world, as an order derived from both the mutual determination of things and its immanent teleological order as a whole, makes clear why Schleiermacher thinks that consciousness of God and consciousness of the world are distinct in their origins but mutually perfecting. The specification of the end of divine love is not necessary for awareness of the world as a world. The mutual determination of things is enough for that. Awareness of the relative freedom, and relative dependence, of things in relation to each other can be known simply through acting on things in the world and being acted upon by them. Likewise, a confused believer could be conscious of the absolute dependence of all things on God, but unaware of the relative dependence of created things on one another. Pious occasionalists, for instance, could be clearly conscious of God but dimly conscious of the world as a world in Schleiermacher's strong sense. Consciousness of God and consciousness of the world are *both* possible, though incomplete, without one another.

Though each is incompletely possible without the other, consciousness of God and world are mutually perfecting: the less the consciousness of God is confused, the keener the distinction between that which is absolutely dependent and that which depends on nothing. And the more determinate this consciousness is rendered, the more the divine love as final cause of the world takes importance. Awareness of the world as ordered and formed for redemption increases in proportion to consciousness of its end. The more consciousness of the world grows, the greater the awareness of its unitary order. And yet, the more one is conscious of the world as a universe, as a single ordered co-existence, the more one is keenly aware that neither the universe as a whole nor anything in it explains the existence of the whole. The more the world is *a world*, the more it is not God. When they grow in concert, consciousness of God and of the world complete each other. The greater the awareness of the divine love as the reason for the world's existence, the more the world is grasped as the totality of finite

232 Spinoza, *Ethics*, IIL7S.

existence created for, and ordered to, the almighty-loving divine *Whence* upon which that order, and so all finite being, is absolutely dependent.[233]

Absolute Miracles

With his account of the nature system as the totality of secondary causes, Schleiermacher is also equipped to give an account of miracles in that system. The mutually-perfecting consciousness of God and world not only yields this notion of the nature system, but the distinction between primary and secondary causes, uncreated and created things, further allows Schleiermacher to clarify the difference between proper miracles and so-called miracles—miracles of another kind.

Given the disputes and confusion in theology and science scholarship over the concept of a miracle, this by itself is informative. At the same time, this distinction solves a controversial exegetical puzzle in Schleiermacher's dogmatics: if, and if so, in what sense, does Schleiermacher admit of the miraculous? There is, of course, general agreement in the literature that Schleiermacher rejects miracles—or at least what he calls an "absolute miracle" [*absolute Wunder*].[234] But in spite of this agreement, there is little clarity about what an "absolute miracle" means, or about why Schleiermacher rejects one kind of miracle but in doing so seemingly implies another kind which he does not reject.[235] Further, there is no scholarly treatment that explicitly discusses the origins of the distinction between absolute miracles and miracles of another sort. Clarity in this distinction is crucial not only for a conceptually rigorous account of the nature system but also to secure coherence and consistency in Schleiermacher's account of divine action. If Schleiermacher admits of *some* genuine miracles, in particular

---

233 "In the divine causality there is no division or opposition anywhere, nor can we regard the government of the world as other than a unity, directed towards a single goal [*als Eine auf Eines gerichtet*]. Hence the Church, or the Kingdom of God in its whole extent as well as in the whole course of its development, forms the one object of the divine world government; whereas any particular thing is such an object only in the other and for it." And further: "These words [predestination and fore-ordination] express far more clearly the relation of each single part to the connected whole, and represent the divine rule of the world as an inwardly coherent order." *GL* §164.3; *The Christian Faith*, 725.

234 "For the more they try to definitely fix an absolute miracle [...]," etc. *GL* §47.2; *The Christian Faith*, 181. The denial of an absolute miracle is correlated to Schleiermacher's denial of the "absolutely supernatural" in general, and the "absolutely supernatural" appearance of the redeemer in particular. See *GL* §13; *The Christian Faith*, 62; and *GL* §47.3; *The Christian Faith*, 183.

235 Sherman, *The Shift to Modernity*, 143.

the coming of Christ as an instance of one, it alters his entire program. If, however, different kinds of miracles can be distinguished, and for good reasons, then Schleiermacher's general account of divine action and the supposed exceptions do not contradict each other, but are instead mutually supporting accounts.

The distinction between absolute miracles and what will be called "relative miracles" is crucial for Schleiermacher's account for a number of reasons. But, in order to understand what Schleiermacher means by this distinction, one must first look to Leibniz. By examining the context in which Leibniz deploys the distinction between absolute and relative miracles it will become apparent what the distinction is meant to uphold and what it is meant to discard. Namely, for both Schleiermacher and Leibniz, an absolute miracle is reserved for that which is absolutely beyond nature itself, and a relative miracle for that which is beyond the power of individual things within nature, but strictly within the power of nature as a whole.

To see this distinction in action we once again turn to *The Leibniz-Clarke Correspondence*. There, in his debate with Clarke, Leibniz introduces a distinction between two kinds of miracles, which he calls miracles of a higher and lower order. "There are miracles of an inferior sort," Leibniz explains, "which an angel can work. He can, for instance, make a man walk upon the water without sinking. But there are miracles, which none but God can work; they exceeding all natural powers. Of which kind are creating and annihilating."[236] The former kind of miracle is uncontroversial. "There is no difficulty among divines," says Leibniz,

> about the miracles of angels. The question is only about the use of that word. It may be said that angels work miracles; but less properly so called, or of an inferior order. To dispute about this would be a mere question about a word. It may be said that the angel who carried Habakkuk through the air, and who troubled the water of the pool of Bethesda, worked a miracle.[237]

Such a miracle can easily be admitted: a miracle of this sort is uncontroversial because it is trivial. If the existence of angels is granted, and if they are beings of a certain kind, a kind which admits of powers such as these, then to carry Habakkuk would be nothing more than an angel acting by its natural powers. It would be equivalent to any other animal doing the things it can do, however powerful, and however extraordinary, in comparison to human capability. Nor is this kind of action any more miraculous given the nature of the object of

236 Leibniz's Fourth Paper, in Alexander, *The Leibniz-Clarke Correspondence*, 43.
237 Leibniz's Fifth Paper, in Alexander, *The Leibniz-Clarke Correspondence*, 93.

the action (Habakkuk in this case). It is, of course, beyond *Habakkuk's* nature to fly, but only as much as any other "violent" action (in the Aristotelian sense). It is no more miraculous for Habakkuk to fly by the power of an angel than for a stone to be lifted by a child. These are only miracles of an inferior sort because they can be entirely explained by the natures of created things.

However, it might be objected that it is much less common for prophets to fly by the powers of angels than for rocks to fly by the powers of children. That is, miracles of this inferior sort might not be as trivial as Leibniz thinks because of their *rarity.* But Leibniz has a reply to this also. "The nature of a miracle does not at all consist in usualness or unusualness," he says, "for then monsters would be miracles."[238] No, "'tis not usualness or unusualness, that makes a miracle properly so called, or a miracle of the highest sort; but it's surpassing the powers of creatures [...]."[239] Unusualness is neither a necessary nor a sufficient condition of the miraculous of the higher order. There are many natural things which are very unusual. But this too is trivial. If one, like Leibniz, holds to the identity of indiscernibles, each and every thing will in some sense be absolutely unique. For if there were two things wholly alike, they would be one and the same thing. Everything to some degree must be unusual. Even apart from this technical sense, there are many strange and rare things which are nevertheless explicable in terms of the powers of created things. And sometimes, as with monsters, rare things are defects. Unusual things are not necessarily unnatural, nor are they necessarily good. Surely, it cannot be right to ascribe every unusual defect to a miracle of God simply because it is rare. And so the unusual can only be called a "miracle" in this qualified sense. As extraordinary as these things might be, they do not surpass the power of created things but can be completely explained by the natural world.

Miracles of the higher order are different. "In good philosophy, and sound theology," says Leibniz,

> we ought to distinguish between what is explicable by the natures and powers of creatures, and what is explicable only by the powers of the infinite substance. We ought to make an infinite difference between the operation of God, which goes beyond the extent of natural powers; and the operations of things that follow the law that God has given them, and which he has enabled them to follow by their natural powers, though not without his assistance.[240]

238 Leibniz's Fourth Paper, in Alexander, *The Leibniz-Clarke Correspondence*, 43.
239 Leibniz's Fifth Paper, in Alexander, *The Leibniz-Clarke Correspondence*, 90–91.
240 Leibniz's Fifth Paper, in Alexander, *The Leibniz-Clarke Correspondence*, 92.

As Leibniz argues, the first step in an account of a miracle proper is to distinguish between primary and secondary, universal and individual, causes. Miracles of the inferior sort are completely explicable in terms of created natures, individual or secondary causes, alone. They are the effects of created things. Miracles of the higher order are not so. They go beyond the power of any and all created things. They are not explicable by any one thing or collection of things within nature, nor by nature as a whole. They can only be explained by that which is beyond the power of all created nature, the power of God. The miracles of a higher order that Leibniz *opposes* then are acts of God upon or within the world. They are actions of a universal cause on an individual cause, or, which is the same, of a primary cause upon a secondary cause. And it is this kind of miracle, a *proper* miracle, which Leibniz does not admit on the basis of the principle of sufficient reason applied to the divine wisdom (see Chapter 3) since the reason for secondary causes is, first and foremost, to produce the intended effects of the primary cause. Any direct divine action implies that secondary causes are insufficient or unnecessary. What does not befit the divine wisdom, God does not do. And so there is no such thing as an absolute miracle *within* nature. The only absolute miracles are, as Leibniz says to Clarke, creation and annihilation: that is, the creation or destruction of secondary causes endowed with specific natures and genuine powers by which to act.

Schleiermacher's distinction between an absolute miracle and miracles of another sort follows this exact pattern. He too distinguishes a stronger, narrower sense of miracle, which he defines as, "simply phenomena in the realm of physical nature which are supposed not to have been caused in a natural manner."[241] This is what Schleiermacher calls an "absolute miracle."[242] Miracles of this sort abrogate the whole system of nature,[243] and it is in regard to miracles of this sort that Schleiermacher warns that, "It can never be necessary in the interest of religion so to interpret a fact that its dependence on God absolutely excludes its being conditioned by the system of Nature."[244] Schleiermacher's "absolute" miracles are Leibniz's miracles "of a higher order." They are beyond not only the power of any individual created thing, but of nature as a whole.

Relative miracles, on the other hand, are miraculous in a wholly different sense. For, unlike absolute miracles, they are explicable by the system of nature. I have called them *relative* miracles because, as Schleiermacher says, "they can

241 *GL* §14 postscript; *The Christian Faith*, 71.
242 *GL* §47.2; *The Christian Faith*, 181.
243 *GL* §47; *The Christian Faith*, 181–82.
244 *GL* §47; *The Christian Faith*, 178.

be called miracles only in a relative sense."[245] They include not only the miracles Christ purportedly performed (as in the context of the previous quotation) but also the incarnation itself, which, though the unique revelation of God, can nevertheless be explained as "a natural fact."[246] Of course, relative miracles, like Leibniz's miracles of an inferior order, can certainly go beyond the power of particular created things and even groups of created things. The revelation of Christ as the beginning of a new religious communion, for instance, "can never be explained by the condition of the circle in which it appears and operates; for if it could, it would not be starting point, but would itself be the product of a spiritual process."[247] That is, the appearance of Christ cannot be explained solely by appeal to the group from which he emerged—however great or small—since that would render him a conclusion, not a beginning; a consequence, not a principle; dependent, rather than that upon which his communion depends. "But," Schleiermacher immediately adds,

> though its [a revelation's] existence transcends the nature of the circle in which it appeared, there is no reason why we should not believe that the appearing of such a life is the result of the power of development which resides in our human nature—a power which expresses itself in particular men and particular points according to laws which, if hidden from us, are nevertheless of divine arrangement, in order through these men to help the others forward.[248]

In other words, the appearance of Christ surpassed the nature of his community while in no way surpassing the power of human nature, a nature which itself developed naturally through the powers and forces inherent in the world as a whole. Like the angel carrying Habakkuk, the appearance of the Redeemer was not supernatural. It surpassed only the powers of particular finite things or groups of them, but not nature itself. This explains why Christ can be a miracle in a system which admits of no miracles. His appearance is a *relative* miracle only, which is, strictly speaking, not a miracle at all.

The supposed exceptions to this rule in *The Christian Faith* can all be explained using this distinction between a relative and an absolute miracle, or explained away as a misunderstanding. For example, Schleiermacher seems to suggest at least one absolute miracle when he speaks of a "special divine activity" [*besondere göttliche Thätigkeit*][249] or "special divine impartation" [*besondere göt-*

245 *GL* §14 postscript; *The Christian Faith*, 72.
246 *GL* §13.1; *The Christian Faith*, 64.
247 *GL* §13.1; *The Christian Faith*, 63.
248 *GL* §13.1; *The Christian Faith*, 63.
249 *GL* §79.1; *The Christian Faith*, 325.

tliche Mittheilung][250] in the context of the divine attributes related to sin and grace. However, it is not clear that either of these is contrary to the claim that God never performs absolute miracles. The former, special divine *activity*, is brought up as a hypothetical and swiftly rejected. *If*, Schleiermacher says, we look for such action, "we are moving in the region of the abstract, and should therefore err were we to look for divine activities bearing upon sin purely by itself."[251] On the other hand, a "special divine impartation" (or, communication) need not refer to a distinct mode of divine action at all, but only to the "specific divine impartation which gives to every approach to salvation the character of grace."[252] In fact, as we learn in Schleiermacher's first discussion of the divine attributes, *all* the attributes denote "something special [*besonderes*] in the manner in which the feeling of absolute dependence is to be related to Him [God]." In other words, the special (or "particular," or "distinct") communication in redemption is nevertheless communicated naturally.[253] Whether sin or grace, "The universal co-operation of God is the same in either domain [...]."[254]

In contrast to absolute miracles, extraordinary things are perfectly admissible. But again, for Schleiermacher as for Leibniz, they are trivial.

> In any other context than that of such faith and its realm we may encounter any number of facts which we cannot explain naturally, and yet we never think of miracle, but simply regard the explanation as deferred until we have a more exact knowledge both of the fact in question and of the laws of Nature.[255]

In this sense, many purported miracles are *potentially* admissible—and thus the theory that, "God has prepared miracles in nature itself," is "pure gain."[256] But, by the very admissibility of relative miracles as natural phenomena, those miracles become susceptible to natural explanation. "In this way," says Schleiermacher, "everything—even the most wonderful [*wunderbarste*] thing that happens or has happened—is a problem for scientific research."[257] What appears to be Schleiermacher's concession to advocates of the miraculous turns out to subsume their position into his own:

---

250 *GL* §80.1; *The Christian Faith*, 326.
251 *GL* §79.1; *The Christian Faith*, 325.
252 *GL* §80.1; *The Christian Faith*, 327.
253 *GL* §50; *The Christian Faith*, 194.
254 *GL* §80.1; *The Christian Faith*, 327.
255 *GL* §14 postscript; *The Christian Faith*, 71–72.
256 *GL* §47.2; *The Christian Faith*, 183.
257 *GL* §47.3; *The Christian Faith*, 184.

> On the whole, therefore, as regards the miraculous, the general interests of science, more particularly of natural science, and the interests of religion seem to meet at the same point, i.e. that we should abandon the idea of the absolutely supernatural because no single instance can be known by us, and we are nowhere obliged to recognize it.[258]

The distinction between primary and secondary causality also has further, perhaps unexpected, consequences for Schleiermacher's original distinction between God and world. As above, the world is the whole of the mutually affectable, and God uniquely and wholly unaffected. But once the distinction between primary and secondary causes, and so the difference between absolute and relative miracles, is explained, Schleiermacher's original distinction between God and world receives renewed importance. The whole of the mutually affectable *is* the totality of secondary causes, and a secondary cause *is* a mutually affecting and affectable thing. Primary causality is distinguished from secondary causality precisely in that *it alone* is wholly conditioning and wholly unconditioned. But when a thing enters into the ordered coexistence of the mutually affectable as a particular cause, it becomes part of that order. It is affected by the things in that order just as it affects other things. Or, put differently, when a universal cause enters into the system of individual causes it must do so as an individual cause. Otherwise it would not *enter into* anything at all. But when it does so, it ceases to be distinguished from other individual causes in kind. As Schleiermacher says:

> Thus the divine omnipotence can never in any way enter as a supplement (so to speak) to the natural causes in their sphere; for then it must like them work temporally and spatially; and at one time working so, and then again not so, it would not be self-identical and so would be neither eternal nor omnipresent.[259]

That is, if God enters into that order, an order defined by the mutually affectability of the things which comprise it, then God ceases to be eternal, omnipresent, and most importantly, loses the self-identity which is the *sine qua non* of an *a se* being. In so doing, God becomes a dependent thing, a creature, a member of the nature system. No, the divine omnipotence does not enter into the system of causes. "Rather," says Schleiermacher, "*everything is and becomes altogether by means of the natural order*, so that each takes place through all and

258 *GL* §47.3; *The Christian Faith*, 183.
259 *GL* §54.1; *The Christian Faith*, 212.

all wholly through the divine omnipotence, so that all indivisibly exist through One."[260]

Notice what this means: not only are absolute miracles within creation inadmissible on a number of other grounds (namely divine wisdom and power), but the very idea of an absolute miracle shows itself as incoherent on the supposition of a single *Whence* on which we find ourselves absolutely dependent, and which is distinguished from all other things by the very fact that it alone depends on nothing. As Schleiermacher says,

> But if, on the other hand, we think of God the Creator in any way as limited, and thus in His activity resembling that which should be absolutely dependent on Him, then the feeling expressing this dependence likewise could not be true (since equality and dependence neutralize each other), and thus the finite in that it resembled God would not be absolutely dependent on Him.[261]

If God is in any way limited, God cannot be that upon which we are absolutely dependent. But, when divine activity resembles finite activity, God is, in fact, so limited. Since every special divine act would become part of the *mutually* conditioning causal nexus, God would be correspondingly conditioned, and thus divine activity resembles finite activity. In every absolute miracle, a universal cause becomes an individual cause, a primary cause becomes a secondary cause, an absolute aseity becomes relatively dependent. *In every absolute miracle, the distinction between God and the world vanishes.*

Surprisingly, this turns the tables on advocates of absolute miracles. The divine freedom and power suggested by absolute miracles is often upheld because, among other reasons, it is seen as necessary to distinguish a free, almighty God from the regular course of natural development. It is partially to avoid notions of God acting solely through the world, and so being collapsed into it, that advocates of absolute miracles champion them. Schleiermacher's account of the God-world distinction—traditional in itself and, in any case, implied by the tradition's principles and assumptions—reveals the consequence of absolute miracles to be the exact opposite. It implies that advocates of absolute miracles are, in spite of their own intentions, most mistaken about the God-world relation. It is they, not Schleiermacher, who flirt most dangerously with a reduction of God to a thing among things (that is, to a member of a genus), or of God to nature. Every absolute miracle confuses the distinction between God and world which Schleiermacher's account of the nature system preserves.

260 *GL* §54.1; *The Christian Faith*, 212. Emphasis added.
261 *GL* §40.3; *The Christian Faith*, 152.

As severe as this problem might seem, it is only one step further to the elimination of God altogether. For as soon as God is no longer that upon which we are absolutely dependent (since, on the supposition of absolute miracles, God is that upon which we are relatively free and relatively dependent just like any other finite thing), there is no single *Whence* which explains the whole *and itself* since the divine life, as partially dependent, is now explained partly by itself, partly by other things. Perhaps counterintuitively, and certainly against the sincere intentions of advocates of absolute miracles, their position amounts not only to a confusion of God and the world, but to a point where God can be eliminated altogether since there is no longer anything that fits the description of the necessary and sufficient conditions of deity. With his refusal to admit of absolute miracles, Schleiermacher is able to uphold the distinction between primary and secondary causality, and so the distinction between God and the world. His more traditional opponents are not.

However, defenders of absolute miracles might object that God need not enter into the nature system as a finite cause at all to affect it. And this might well be right. But it was against this view that the previous chapter's argument from wisdom and the following chapter's argument from plenitude are directed. These arguments show divine action of this sort always implies a deficiency between potency and act as regards God's goodness and power respectively. For if absolute miracles do not *enter into* the world, they *intervene upon* it.[262] It is to avoid the latter that theologians might offer the former. But it is the former, the notion of divine action within the world which does not interrupt it because it is somehow a feature of it, which is excluded by a sufficient account of the world as distinct from God.

But still, the objection might be pressed, why cannot God act within the world, within the system of mutually affectable causes while nevertheless also remaining *a se* and so genuinely God? This objection might seem more promising: it simply offers a "both-and" solution. But it nevertheless fails on simple grounds. The reason God cannot be a universal and an individual cause at once is simply that individual/secondary causes and universal/primary causes are mutually exclusive categories. And *it is their very exclusivity* that is the basis of a creator-creature distinction. For a universal cause is *uniquely* the reason for its own existence, that upon which everything depends and which itself

---

262 I offer this distinction, not to defend the distinction myself, but so that advocates of absolute miracles might see how the difficulties Schleiermacher presses upon them apply regardless of the metaphor employed. Whether miracles are viewed as breaking the causal chain, or as taking place within it, different, though equally pressing, problems result. See Chapter 5 for Schleiermacher's argument against God altering the determinate chain of causes.

depends on nothing. And individual causes are every other thing, every creature, distinguished from the universal cause by their dependence on it. Unlike God, individual causes are not wholly the reason for their own existence. They depend partially on other finite things and absolutely on God to be what they are. To say that a universal cause could be an individual cause is to say that which depends on nothing could depend on something, which is absurd. And if it is nevertheless insisted upon that this absurdity is a necessary theological paradox, Schleiermacher's claim holds no less well: the champion of absolute miracles is obliged to defend the dissolution of the creator-creature distinction because, in the final analysis, objections to Schleiermacher's argument can succeed only on the condition that they equivocate between primary and secondary causes, between God and world—which is precisely Schleiermacher's point.

Conclusion

To explain that a world is the whole of the mutually affectable, ordered as thing to thing and as all things to God, explains why Schleiermacher thinks there is a nature system in his strong sense. The world simply is this totality, uniquely and perfectly created, and formed for the end of redemption. Further, the indexing of two different powers of consciousness (of God and of the world) explain why Schleiermacher thinks each perfects the other in piety while at the same time strengthening, rather than diminishing, the distinction between God and the world. The difference between a universal and an individual cause (between primary and secondary causes, as the tradition knows them) explains the difference between God and the world, while nevertheless preserving the absolute dependence of all things on God. This leads him to follow Leibniz in distinguishing between absolute and relative miracles in order to preserve the distinction between universal and individual causes, between God and the world. The result is not only an account of the world as a world, and the world as distinct from God, but by this very distinction Schleiermacher sets up another argument against absolute miracles. God is distinct from the world precisely to the degree that divine causality can be distinguished from natural causality. When the latter distinction is collapsed, so is the former. Ironically, in light of the history of attacks against him, it is Schleiermacher who, unlike his opponents, not only values, but actually sustains a sufficient creator-creature distinction.

One of the classic objections to Schleiermacher's account of divine action is that Schleiermacher's case relies too heavily on an account of the nature system as an assumption. But as the indexing of the God-world distinction to the universal cause-individual cause distinction demonstrates, the nature system is not an assumption, let alone an unexamined one. Instead, like all of the arguments in support of the eternal covenant, it is grounded in his doctrine of God. In the case of the world, all that is added to the principle of divine uniqueness and independence is the empirical premise that there is in fact a world, and that, as a matter of experience, at least some things in it are the genuine causes of their own effects. Critics of the nature system must be prepared to deny these basic beliefs. But Schleiermacher's account of the nature system follows from such plausible premises that attempts to resist Schleiermacher's case by denying the nature system are far more suspect than Schleiermacher's theological conclusions.

However, Schleiermacher's account of the nature system is not only another step in his case against absolute miracles. Besides being crucial for his criticism of absolute miracles as an account of divine action—criticism which we will explore further in the following chapter—Schleiermacher's account is also an important part of what will be his alternative account, the basis of the eternal covenant. Not only does good natural science depend on an account of the world as a world, but so does good theology. As Schleiermacher will argue, if miracles obscure knowledge of God, the order of the world makes the divine wisdom and love plain, as we will see in Chapter 6.

Chapter 5 Divine Power and the Necessity of the World: Spinoza and Schleiermacher on the Perfection of Nature

> "Are we right in saying that there is one world, or that there are many and infinite? There must be one only if the created copy is to accord with the original."
>
> - Plato, *Timaeus*

For all the strengths of Leibniz's argument that God must, by virtue of his goodness, act for the best, Clarke was not satisfied. In God's inevitable choice of the best he saw the destruction of divine freedom. As he told Leibniz,

> This argument, if it were good, would prove that whatever God can do he cannot but do; and consequently that he cannot but make everything infinite and everything eternal. Which is making him no governor at all, but a mere necessary agent, that is, indeed no agent at all, but mere fate and nature and necessity.[263]

Leibniz shared Clarke's sense of the danger of reducing God to "mere fate and nature and necessity," and yet he was not persuaded by Clarke's reply. Instead he responded that Clarke "often endeavors to impute to me necessity and fatality [...] But he says so without proving it and without taking notice of the explications I have formerly given, in order to remove the difficulties that may be raised upon that head."[264] The distinctions Leibniz offered were crucial to his position, and it was these distinctions which culminated in his famous doctrine of the best of all possible worlds. His rejoinder to Clarke's *Reply* was a recapitulation of this doctrine.

Schleiermacher, however, does not follow Leibniz down this road. Instead, he not only admits, but argues for, the conclusion that God does what God does by nature and necessity—though utterly denies that this is *mere* fate. Following Spinoza, Schleiermacher argues that positions like Leibniz's are untenable, and that they inevitably reduce to his own. But unlike Clarke, and again following Spinoza, he presses on his opponents that it is not his view which is impious, but theirs. It is only if God acts by the necessity of the divine nature that God is completely free, and it is only if God does all that an infinite intellect

263 Clarke's Fourth Reply, in Alexander, *The Leibniz-Clarke Correspondence*, 50.
264 Leibniz's Fifth Paper, in Alexander, *The Leibniz-Clarke Correspondence*, 55–56.

DOI 10.1515/9783110542301-005

can conceive that God is almighty. And while Schleiermacher does not give up the teleological nature of the universe, he denies that the good for which God acts, and the means by which God does so, are possibilities amongst other possibilities—even a possibility which God inevitably chooses.

Moreover, Spinoza and Schleiermacher apply the notion that the world necessarily is such as it is as another argument against supernatural emendation or supplementation. This time, however, the argument does not hinge on wisdom, but on necessity. Since everything that happens does so as determinate effect from determinate cause, and since both argue that that determinate cause is the divine *essence*, they conclude that things cannot be otherwise without contradicting the divine essence. This time the perfection under consideration is not more or less fitting means to an end, but rather, this perfection is more or less in regard to metaphysical possibility. For Spinoza and Schleiermacher there is no remainder to what God *may* do, since God does all that is possible for God to do by the necessity of the divine nature alone. Since God cannot *be* otherwise, God cannot *do* otherwise. Since it is by the necessity of the divine essence that the natural order is as it is, God can never supplement or emend that order, for the divine nature cannot contradict itself. In consequence, as with Leibniz's argument in the previous chapter, there is nothing which is not explicable *in principle* through the powers and forces inherent in the natural order. As with Leibniz's account before, Schleiermacher uses Spinoza's account of God to fund his account of nature.

The aims of this chapter are: first, to explain the important technical distinction between hypothetical and absolute necessity that underlies Leibniz's disagreement with Spinoza; two, to give Spinoza's rejoinder to accounts like Leibniz's; and three, to show how, on this critical point Schleiermacher sides with Spinoza against Leibniz and holds to the notion of God acting of *absolute* necessity. Only on the basis of this technical distinction will Schleiermacher's full account and its implications become clear.

Leibniz on the Hypothetical Necessity of the World

When Clarke and Leibniz accused one another of propounding a position which reduced to mere fate and nature and necessity, each understood the other's position as resulting in a denial of God's agency.[265] For his part, Clarke saw that

265 Leibniz's Fourth Paper, Clarke's Fourth Reply in Alexander, *The Leibniz-Clarke Correspondence*, 39, 50.

Leibniz's position carried its own necessity: the necessity of the best. Since the best is known by God as *eternally* best, that means that there will never be another best to will; and since God cannot but will the best, that implies the eternal necessity of the best. And that seems to approach the "fate and nature and necessity" which Clarke was so concerned to deny. On the other hand, Leibniz saw in Clarke's position the same mere fate but in mirror image. Since, for Clarke, a sufficient reason for God's willing a thing was simply the will of God apart from its reference to the good willed, God's willing appeared more fatalistic than any necessity of the good.[266] For Clarke, God's will was free insofar as it was not tied to a particular way of being put into execution. But, the failure to tie the will of God to an ordained end meant that the will of God need not act for a determinate *good* but is, instead, its own sufficient reason for acting, apart from a determinate end for which to act. The result was a dilemma: God's agency was impoverished if divine activity was not rendered determinate, and so necessary, by a good for which to act and a means by which to act best; or, God's activity, curtailed by no extrinsic good and no best means, was correspondingly indeterminate. One way made God subject to something other than Godself; the other way made the divine will accountable to nothing at all, not even the divine nature. In each case, it was the perceived "fate and nature and necessity," exemplified by Spinoza, which each sought to avoid.

For his part, Leibniz was satisfied that he had avoided Spinoza's reduction to natural necessity (see below) through his introduction—or, rather, *re*-introduction—of a distinction between different *kinds* of necessity. He proposed two such distinctions: the distinction between metaphysical necessity and moral necessity, and between absolute necessity and hypothetical necessity. These distinctions are crucial for understanding Leibniz's position and its alternative in Spinoza and Schleiermacher. As Leibniz explained, "There are necessities which ought to be admitted. For we must distinguish between an absolute and an hypothetical necessity."[267] He went on to specify that

> We must also distinguish a necessity which takes place because the opposite implies a contradiction; (which necessity is called logical, metaphysical, or mathematical;) and a necessity which is moral, whereby, a wise being chooses the best, and every mind follows the strongest inclination.[268]

266 Leibniz's Third Paper in Alexander, *The Leibniz-Clarke Correspondence*, 27–28.
267 Leibniz's Fifth Paper in Alexander, *The Leibniz-Clarke Correspondence*, 56.
268 Alexander, *The Leibniz-Clarke Correspondence*, 56.

At this point, Leibniz meant to distinguish *both* absolute from hypothetical necessity *and* metaphysical from moral necessity. He seemed to believe that the distinction between absolute and hypothetical necessity is a necessary condition for the distinction between metaphysical and moral necessity, or even for the maintenance of moral necessity at all. We will return to this in the following chapter to see how Schleiermacher argues that, in fact, no such distinction is necessary. But for now it is worth reiterating that even for Leibniz the question of moral necessity is related to, but distinct from, the question of absolute or hypothetical necessity.

The distinction between absolute necessity and hypothetical necessity is this: "Hypothetical necessity is that," Leibniz said, "which the supposition or hypothesis of God's foresight and pre-ordination lays upon future contingents."[269] And, Leibniz rightly claimed that this necessity at least must be admitted if divine foreknowledge is not to be denied. And yet Leibniz claimed this necessity is not absolute since its opposite does not imply a contradiction—the condition of an absolute necessity. Both are necessary, but the one is necessary only on the condition of a hypothesis (or supposition, which is the same thing). The other is necessary because its opposite implies a contradiction.

To understand this distinction better, it is helpful to turn to its use in a long tradition predating Leibniz. An early use is found in Boethius's *Consolation of Philosophy*, where the two kinds of necessity are called "simple" and "conditioned."[270] "So, then," Boethius says, "there are two necessities—one simple, as that men are necessarily mortal; the other conditioned, as that, if you know that someone is walking, he must necessarily be walking."[271] The former simple necessity is that which is imposed by the nature of a thing, and its contrary, is impossible. For example, by definition all men are necessarily mortal. The latter, conditioned necessity, is synonymous with Leibniz's hypothetical necessity. "For," says Boethius, "the former necessity is not imposed by the thing's own proper nature, but by the addition of a condition. No necessity compels one who is voluntarily walking to go forward, although it is necessary for him to go forward at the moment of walking."[272] That is, it is perfectly possible to be a man, but not be presently walking. *When* a man is walking he must necessarily be walking, but when he ceases to walk he does not thereby cease to be a man. His walking is accidental to his nature. As such, there is no contradiction implied

---

269 Alexander, *The Leibniz-Clarke Correspondence*, 56.
270 Boethius, *Consolation of Philosophy*, trans. H. R. James (London: Eliot Stock, 1897), v. 6.
271 Boethius, *Consolation of Philosophy*, v. 6.
272 Boethius, *Consolation of Philosophy*, v. 6.

in either case and, therefore, this kind of necessity is not absolute/simple—hence Leibniz's distinction.

Leibniz also introduced another kind of necessity: moral necessity. Moral necessity is that necessity by which a wise being chooses the best. "But good, either true or apparent; in a word, the motive, inclines without necessitating; that is, without imposing an absolute necessity."[273] Moral necessity is, like hypothetical necessity, distinct from absolute necessity. And it is distinct in exactly the same way: namely, that its opposite does not imply a contradiction. "For," Leibniz says, "when God (for instance) chooses the best; what he does not choose, and is inferior in perfection, is nevertheless possible. But if what he chooses, was absolutely necessary; any other way would be impossible: which is against the hypothesis. For God chooses among possibles, that is, among many ways, none of which implies a contradiction."[274] All moral necessities are kinds of hypothetical necessity, but not all hypothetical necessities are moral necessities. Assuming the choice of best is an *actual* choice, that is, a choice which has actually been made, that choice becomes hypothetically necessary. Just as a man must hypothetically be walking if, in fact, he is walking, so a wise being must hypothetically choose the best if, in fact, he has done so. This need *not* imply that the wise might not choose the best.[275] After all, moral necessity is still a kind of necessity. But, having chosen the best according to wisdom, the best chosen acquires this second kind of necessity—hypothetical necessity—so that if God wills, say, to create the best of all possible worlds, God must necessarily do so, so long as God has, in fact, so willed.

And yet, even as these necessities build up, none of them makes *this* world (the best which God has, in fact, chosen) *absolutely* necessary since, according to Leibniz, none of the other possible worlds' existence implies a contradiction either. Leibniz claimed to have reached this conclusion *ex hypothesi* since the possibility of worlds demands that their existence does not imply a contradiction—otherwise they were not possible: "But if what he chooses, was absolutely necessary; any other way would be impossible: which is against the hypothesis."[276] Unfortunately, as we will see, in the face of Spinoza's and Schleiermacher's criticisms he appears to have begged the question. But notice the *supporting* role Leibniz's doctrine of possible worlds played in maintaining the distinction between hypothetical/moral and absolute necessity. His distinction between kinds of necessity was not ordered to supporting his doctrine of possible worlds,

---

273 Leibniz's Fifth Paper in Alexander, *The Leibniz-Clarke Correspondence*, 57.
274 Leibniz's Fifth Paper in Alexander, *The Leibniz-Clarke Correspondence*, 57.
275 See: Garber, *Leibniz: Body, Substance, Monad*, 236.
276 Alexander, *The Leibniz-Clarke Correspondence*, 57.

but rather the other way around: it is the doctrine of possible worlds which kept this distinction up and running since, if the concept of a possible world included the predicate "which does not and will not ever exist," then their existence *would* imply a contradiction. And, of course, that would make the existence of this world, and only this world, absolutely necessary. Since Leibniz meant to maintain the hypothetical, moral, and absolute necessity distinctions, he must hold that no contradiction is implied by the existence (or not) of possible worlds, and therefore that they are in some sense *genuinely* possible.[277]

Spinoza on the Absolute Necessity of the World

Unlike Leibniz, Spinoza held that the world followed from God by absolute (or, in Spinoza's words, "geometrical" or "natural") necessity. His reasons in the *Ethics* all rely on the divine perfection and the removal of limits. Two distinct appeals are made which dovetail in support of his argument: one appeal is to the perfection of divine power; the second to the perfection of divine freedom —both of which ultimately amount to the same thing. In consequence, everything that happens does so as a determinate effect issuing necessarily from a determinate cause such that not only does everything in this world happen in one way and no other, but that this world, and this world alone, is what God creates.

Spinoza's first appeal in support of his argument is to the perfection of divine power. That is, "From the necessity of the divine nature there must follow infinitely many things in infinitely many modes, (i.e., everything which can fall under an infinite intellect)."[278] In other words, Spinoza subscribes to the principle of plenitude: that God does all that God can do. Since God can do no more, God can do no other. But why can God do no more? For the answer to that question, Spinoza appeals to divine freedom. God's perfect power is a consequence of God's perfect freedom.

Spinoza's analysis of that freedom begins in *Ethics I* with a consideration of *substance*. As D3 states, "By substance I understand that which is in itself and conceived through itself, that is, that whose concept does not require the concept

---

277 This is a reason to qualify Michael Griffin's reading of Leibniz's hypothetical necessity as amounting to the same logical status as Spinoza's geometrical necessity. Both Leibniz and Spinoza treat geometrical/absolute necessities as those whose opposites imply a contradiction, but in the *Leibniz-Clarke Correspondence* Leibniz reiterates that the contrary of a hypothetical necessity does *not* imply a contradiction. That is, the two necessities have very different modal force. See Griffin, *Leibniz, God and Necessity* (Cambridge: Cambridge University Press, 2013), 65, 73.

278 Spinoza, *Ethics*, IP16.

of another thing, from which it must be formed."[279] Spinoza takes it to be axiomatic that some things require explaining through other things. That is, by A1, he tells us, "Whatever is, is either in itself or in another."[280] And lest the explicability of things fall into infinite regress, he also finds it axiomatic that "What cannot be conceived through another, must be conceived through itself."[281] Since, by D5, there are at least some things which cannot be explained through themselves alone,[282] recourse will have to be had to other things in order to explain those modes. And either those things will explain themselves or not. If not, the explanation continues until its eventual termination in a substance, "that which is in itself and conceived through itself," since everything must be either conceived through another or through itself. Therefore there must be at least one substance.

By D6, we are told that God is such a substance, "a being absolutely infinite, that is, a substance consisting of an infinity of attributes, of which each one expresses an eternal and infinite essence."[283] To the uninitiated, that might seem to imply that substance is a class term, a genus. Instead this is exactly what Spinoza intends to deny. For by P5, "two or more substances of the same nature or attribute"[284] cannot exist: if they were truly two they would have to be distinguished *by their attributes* (that is, by D4, "by what the intellect perceives of a substance as constituting its essence"[285]), and so, would not be the same. But, if the same, then they would not differ at all and, being the same in every way, so really be, by a tacit application of the principle of the identity of indiscernibles, one substance not two.[286] Further, by P8, "Every substance is necessarily infinite." That is, by D2, no substance is limited by another of the same nature. For if limited by another of the same nature, that other nature would have to share an attribute in common with the first nature. "And so there would be two substances of the same attribute, which is absurd (by P5)."[287] Therefore, there is one and only one substance, which, by P8, is necessarily infinite.[288] Therefore, "God, or a substance consisting of infinite attributes, each of which expresses eternal and infinite essence, necessarily exists."[289]

---

279 Spinoza, *Ethics*, ID3.
280 Spinoza, *Ethics*, IA1.
281 Spinoza, *Ethics*, IA2.
282 Spinoza, *Ethics*, ID5.
283 Spinoza, *Ethics*, ID6.
284 Spinoza, *Ethics*, IP5.
285 Spinoza, *Ethics*, ID4.
286 Spinoza, *Ethics*, IP4, P5D.
287 Spinoza, *Ethics*, IP8D.
288 Spinoza, *Ethics*, IP8S1,2.
289 Spinoza, *Ethics*, IP11.

Because God alone is absolutely infinite in nature, nothing can hinder God. In order to hinder God it must, by D2, share some attribute with him.[290] Otherwise, by P2, those two things—God and the thing which would hinder God—would have nothing in common with one another.[291] So, whatever would limit God must share an attribute with him. But, as above, it is an absurdity to conceive of two substances of the same attribute since a substance is that "which is in itself and is conceived through itself alone."[292] Therefore, since God is absolutely unhindered, God must be and act through the divine nature alone. And since that which "exists from the necessity of its nature alone, and is determined to act by itself alone" is free,[293] God alone is free, and supremely so. Absolutely unhindered, God is unlimited in God's nature, that is, by D6,[294] God is absolutely infinite. And since God is and acts by the necessity of the divine nature alone, "From the necessity of the divine nature there must follow infinitely many things in infinitely many modes [...]."[295]

The payoff to this line of argument comes in the following extended proposition, P17. Here Spinoza concludes that since there is nothing outside God which could determine God to act (since all things but substance are modes of substance, and there is only one substance) that, "God acts from the laws of his nature alone, and is compelled by no one, q.e.d."[296] And, "From this it follows, first, that there is no cause, either extrinsically or intrinsically, which prompts God to action except the perfection of his nature."[297] God therefore acts by the laws of his perfect nature alone, and those laws and that nature are one and the same.

Only in the scholium to P17 is the target of this buildup of propositions made plain. Here we learn that his opponents—who remain anonymous—hold two views, each of which Spinoza intends to deny. First, they "think that God is a free cause because he can (so they think) bring it about that the things which we have said follow from his nature (i.e. which are in his power) do not happen or are not produced by him."[298] That is, at least some of those thinkers that Spinoza opposes hold that God has the power both to will and to nil, to bring about

---

290 Spinoza, *Ethics*, ID2.
291 Spinoza, *Ethics*, IP2.
292 Spinoza, *Ethics*, ID3
293 Spinoza, *Ethics*, ID7.
294 Spinoza, *Ethics*, ID6.
295 Spinoza, *Ethics*, IP16.
296 Spinoza, *Ethics*, IP17D.
297 Spinoza, *Ethics*, IP17C.
298 Spinoza, *Ethics*, IP17S1.

some things *and* to bring about their negation. Presumably they do this to uphold the basis of the distinction between hypothetical and absolute necessity: that there are at least some things that exist, the non-existence of which does not imply a contradiction.

Second, those opponents,

> [T]hough they conceive God to actually understand in the highest degree, they still do not believe that he can bring it about that all the things he actually understands exist. For they think in that way they would destroy God's power. If he had created all the things in his intellect (they say), then he would have been able to create nothing more, which they believe to be incompatible with God's omnipotence.[299]

That is, his opponents hold that the possible in God's intellect extends beyond the actual, that God does *not* do all that God can. Spinoza's opponents deny the principle of plenitude.

In reply, Spinoza accuses his opponents (again, unnamed) of denying divine omnipotence, and so ultimately God's perfection:

> For they are forced to confess that God understands infinitely many creatable things, which nevertheless he will never be able to create. For otherwise, if he created everything he understood he would (according to them) exhaust his omnipotence and render himself imperfect. Therefore to maintain that God is perfect, they are driven to maintain at the same time that he cannot bring about everything to which his power extends.[300]

That is, according to Spinoza's opponents, God understands infinitely many things which do not exist but whose existence does not involve a contradiction, and yet which nevertheless *cannot* exist, which is absurd: for if those things existed, then they would be actual and not merely possible. And since it is his opponents' express intent to deny that everything which is possible for God becomes actual, they *must* contend that there are at least some things whose existence does not involve a contradiction, and yet which cannot exist.

Of course, those who would maintain such a view, like Leibniz, would distinguish God's intellect from his will, so that God may know something without thereby willing it. For his part, Spinoza intends to deny an attribute of "will" to God altogether.[301] "Nevertheless," says Spinoza, "to please them, I shall show that even if it is conceded that will pertains to God's essence, it still follows from God's perfection that things could have been created by God in no other

299 Spinoza, *Ethics*, IP17S1.
300 Spinoza, *Ethics*, IP17S1.
301 Spinoza, *Ethics*, IP17S2.

way or order."[302] On traditional grounds, Spinoza's argument from perfection applies equally well.

First, as even Spinoza's opponents concede, "it depends on God's decree and will alone that each thing is what it is. For otherwise God would not be the cause of all things."[303] God must be the cause of all things because there is only one substance, one thing which is and acts by itself alone. Since it is axiomatic that all things must either be conceived through themselves or through another, all can only be conceived through that which is and acts through itself alone: a substance of infinite attributes, that is, God. Everything ultimately depends on God's will alone to be what it is.

But that is not all. Things could have been in no other way or order since, "all God's decrees have been established by himself from eternity. For otherwise he would be convicted of imperfection and inconstancy. But since, in eternity, there is neither *when*, nor *before*, nor *after*, it follows, from God's perfection alone, that he can never decree anything different, and never could have, *or* that God was not before his decrees and cannot be without them."[304] That is, since God's essence involves existence, God is eternal.[305] Since God's attributes express what pertains to substance, God's attributes must be eternal. And since God's existence and essence are one and the same, God is immutable; and since the divine attributes express the divine essence, the divine attributes are immutable.[306] Because God's decrees are decrees of the divine will, and the divine will is immutable, the divine decrees cannot be otherwise. Indeed, Spinoza's opponents, should they subscribe to the divine aseity—and the simplicity, eternity, and immutability that follow from it—would be obliged to concede that things could not have been in any other way or order than they, in fact, are.

Yet his opponents would object to this conclusion: "But they will say that even if it were supposed that God had made another nature of things, or that from eternity he had decreed something else concerning Nature and its order, no imperfection in God would follow from that."[307] That is, they would wish to claim that God's *will* is hypothetically necessary such that having willed this particular natural order, God cannot will otherwise, but *not* that God could will no other nature of absolute necessity.[308] In other words, having grant-

---

302 Spinoza, *Ethics*, IP33S2.
303 Spinoza, *Ethics*, IP33S2.
304 Spinoza, *Ethics*, IP33S2.
305 Spinoza, *Ethics*, ID8.
306 Spinoza, *Ethics*, IP20.
307 Spinoza, *Ethics*, IP33S2.
308 See: Thomas Aquinas, *Summa Theologiae*, Ia, q. 19, a. 3, resp.

ed that given the divine decree the natural order that follows from that decree does so in no other way or order, they must maintain that the decree itself is merely hypothetically necessary so that the concept of another *decree* does not involve a contradiction even if, supposing a decree, another nature and order does involve a contradiction.

This shift is of no avail. On the one hand, "if they should say this, they will concede at the same time that God can change his decrees."[309] If they admit that *if* God had decreed differently, that no imperfection would follow, they are obliged to admit that God could have *actually* decreed differently. That is, his opponents commit themselves to a theory of "possible decrees" on analogy with possible worlds. Having admitted that God's decrees could have been otherwise, they thereby admit that they are hypothetically changeable. And having admitted that, "why," Spinoza presses, "can he [God] not *now* change his decrees concerning created things, and nevertheless remain equally perfect?"[310] In other words, if change is admitted in regards to God's decrees at all, they are, in principle, changeable at any moment.

The threat of a changeable decree, though distasteful, was minor compared to what followed. Because Spinoza and his opponents all subscribed to the divine simplicity, they all admitted that God's "intellect and will are not distinguished from his essence."[311] In other words, they concede D4, that an attribute is "what the intellect perceives of a substance, as constituting its essence."[312] But since the decree is decreed by the will of God, having conceded that God's attributes are essential, "it follows that if God had another actual intellect, and another will, his essence also would necessarily be other. And, therefore (as I inferred at the beginning), if things had been produced by God otherwise than they now are, God's intellect and his will, that is (as conceded) his essence, would have to be different. And this is absurd."[313] That is, since God's decree *is* God's will, if God's decree were different, God's will would be different. But since God's will constitutes God's *essence*, if God's will were different God's essence would be different. Therefore, God's decree cannot be different without changing God's essence. Since, by P19D and P20C2 God's essence—and so God's attributes—are eternal and immutable, the possibility of a different decree would be absurd.

The effort to avoid the principle of plenitude—that God does all that God can do—by holding to the hypothetical necessity of God's will does not, according to

309 Spinoza, *Ethics*, IP33S2.
310 Spinoza, *Ethics*, IP33S2. Emphasis added.
311 Spinoza, *Ethics*, IP33S2.
312 Spinoza, *Ethics*, ID4.
313 Spinoza, *Ethics*, IP33S2.

Spinoza, succeed. And, in fact, it cannot succeed since God's will is essential, and therefore eternal and immutable. If God could will differently, God could *be* differently. But God's existence is necessary, such that the divine nature necessarily is such as it is. And, therefore, God's will is as necessary as God's existence, since both are essential. Should one seek to avoid the necessity of the divine will by distinguishing it from the divine essence, one would run afoul of the self-subsistence and uniqueness of substance, as above. In order to avoid polytheism (multiple substances), Spinoza claims that God's will must be essential. And, since essential, it must be one and the same with the divine intellect (which is also essential) so that whatever God knows God wills.

Because all that God can do God does, what God does not do cannot be done. That is, what God does not do is impossible. Likewise, since God's decree cannot be otherwise, everything not only does, but *must* happen in no other order or way. Just as what God cannot do is impossible, so what God does is necessary. In both cases, that means that the existence of impossible worlds and the non-existence of the actual world implies a contradiction: their respective concepts *do* contain their existence or non-existence since, as modes, it is only *through substance* that they can be conceived.[314] Moreover, since, if a plurality of Gods is to be avoided, the existence or non-existence of worlds *must* be explained through the divine essence and not through themselves. Otherwise, if conceived through themselves, they would be other substances, which by P14 —that "Except God, no substance can be or be conceived"—is impossible.[315] In consequence, the predicated existence or non-existence of worlds must be predicated in relation to God and never in isolation, since it is by the divine essence alone that anything exists. Since, in relation to the divine essence, one and only one world can and does exist, its non-existence implies a contradiction; all so-called possible worlds cannot and will not exist, and so their existence likewise implies a contradiction.

The distinction between absolute and hypothetical necessity relies on at least some worlds, the existence of which does not imply a contradiction, but which nevertheless do not exist. Spinoza's argument, by tying the existence or non-existence of things to the necessity of the divine essence, renders their existence necessary. By doing so, he renders the so-called possible existence of worlds, other than this one, a contradiction. And, in consequence, he reduces the distinction between absolute and hypothetical necessity to natural necessity.

---

314 Spinoza, *Ethics*, ID5; P15.
315 Spinoza, *Ethics*, IP14.

In the final analysis, there is no such thing as hypothetical necessity. Everything happens by the necessity of the divine nature alone.

But, if everything that happens happens by the same necessity of the divine nature, the worry arises that if the world should exist by this necessity, that its essence therefore involves existence—a dignity reserved for God alone. Should Spinoza's position imply this, it would amount to a strong reason to reject his conclusion. Of course, such a conclusion would be a very surprising result of an argument which relies on the uniqueness of God as a *premise*. Even so, the worry remains.

Spinoza has a response to this concern. First, he claims in P24 that "The essence of things produced by God does not involve existence."[316] That is, created things are not their own reason for existence, but depend on God at every moment. It is God alone whose essence involves existence "considered in itself."[317] That is, God is affirmed as that which can uniquely be and be conceived through itself alone.[318] Everything else's existence is dependent, and can neither be, nor be conceived, except through God. This does not negate Spinoza's earlier claim that all things follow from the divine essence of natural necessity. Instead, he distinguishes between two reasons why a thing is necessary[319]: "A thing is called necessary *either* by reason of its essence *or* by reason of its cause. For a thing's existence follows necessarily either from its essence and definition or from a given efficient cause."[320] God's existence alone follows from the necessity of the divine essence. The existence of all dependent things follows instead from the necessity by which a given determinate effect follows of natural necessity from a given determinate cause—a necessity which Spinoza again thinks is simply axiomatic.[321] The world is a necessary effect of God's essence. Given a cause, its effects follow necessarily. But, the necessity by which effects follow from causes does not thereby make effects causes, nor do effects cease to depend on their causes. Instead, the order of dependence of effects on causes is preserved. Just as effects follow necessarily from causes, but are nevertheless dependent on their causes, so the world follows necessarily from the divine essence, but is nevertheless dependent on that essence for its existence.

An example might help to further clarify. Take, for instance, Julius Caesar. Julius Caesar's concept includes his existence, which we know because he ac-

---

316 Spinoza, *Ethics*, IP24.
317 Spinoza, *Ethics*, IP24D.
318 Spinoza, *Ethics*, IPD3.
319 See Griffin, *Leibniz, God and Necessity*, 73.
320 Spinoza, *Ethics*, IP33S1. Emphasis added.
321 Spinoza, *Ethics*, IA3.

tually existed. Since, according to Spinoza, any determinate effect follows *necessarily* from its cause, once we know that Julius Caesar was a part of the world, an effect of God, we know that he followed necessarily from God. In that sense, Julius Caesar necessarily exists. And, in fact, his possible non-existence contradicts his concept since his existence follows necessarily from the divine essence which itself necessarily is such as it is—without which Caesar could neither be *nor be conceived.* At the same time, Julius Caesar's essence does not involve existence in the sense that he exists and is explained through himself alone. He is not the self-existent Julius Caesar, even though he necessarily exists. Instead Caesar's existence must be explained through another: God, whose essence uniquely involves existence. *In no way, however, is Julius Caesar's existence merely hypothetically necessary.* Although Julius Caesar's essence does not involve existence in the sense that Caesar is not *a se*, nothing, including his existence, could have happened in any other order or any other way without ultimately contradicting the *divine essence* from which all things follow as necessary effect from necessary cause. Though the world follows from God of absolute necessity, it does not exist through itself, but only through God.

On Spinoza's grounds there is no possibility of confusing the world's necessity with God's necessity such that the world becomes independent of God or God dependent on the world. But, granted as much, does that not still make the world follow from the divine essence in such a way that God is in some sense subject to fate? Once again the answer is 'no.'

On the one hand, God cannot be subject to fate because fate and freedom are contraries—assuming fate implies a kind of compulsion which its derogatory use suggests. By D7, that which is supremely free is utterly free from compulsion since it is determined to act by itself alone.[322] And, by P17, Spinoza shows that God is the *one and only* thing that is determined to act by his nature *alone.*[323] Such freedom follows necessarily from a consideration of substance since a substance exists and acts through itself alone. Hence, God cannot be subject to fate, but is instead supremely and uniquely free.

On the other hand, not only does Spinoza defend his conception of divine freedom against a tradition that wishes to distinguish God's will from his nature,[324] but he further argues against that tradition that it is, in fact, *their* conception of freedom which subjects God to fate and compulsion. If God uniquely exists and acts through himself alone, then "From this it follows, first, that there is

322 Spinoza, *Ethics*, ID3.
323 Spinoza, *Ethics*, IP17
324 Spinoza, *Ethics*, IP17S2; P32C1, C2; P33S2; App.

no cause, either extrinsically or intrinsically, which prompts God to action, except the perfection of his nature."[325] This follows immediately from a consideration of substance: for if something other than God's nature prompts God to action, then God's existence is composite, and God is not therefore that which exists and is conceived through itself alone. If an *intrinsic* cause like a distinct will is supposed, that will must ultimately be determined to exist and so act by the divine nature on pain of contradiction of the divine aseity. And, if so determined, that will is explained, as all other things, by the necessity of the divine nature. In the same way, if an *extrinsic* cause, like an external end, is proposed, equally disastrous consequences ensue. As Spinoza explains:

> For they seem to place something outside God, which does not depend on God, to which God attends, as a model, in what he does, and at which he aims, as at a certain goal. This is simply to subject God to fate. Nothing more absurd can be maintained about God, whom we have shown to be the first and only free cause, both of the essence of all things, and of their existence.[326]

While this does not rule out *every* possible account of God's will or the good for which God acts (see Chapter 6), it does rule out those accounts which distinguish either from God's nature. If they are non-identical with God's essence, they contradict the divine freedom and the concept of substance itself; but if identical with God's essence, they are simply one and same thing and so God does, after all, act by the necessity of his nature alone.

Miracles and the Necessity of the Natural Order

Spinoza's reduction of hypothetical necessity to natural necessity will later be put to use by Schleiermacher in arguing for the theological benefits of the covenant. But first and foremost, Spinoza took his position to undermine the notion of miracle as supplementary divine action. He held that, as a result of the necessity by which everything follows from the divine essence, no action in the world ever falls outside the order of natural causes. Since determinate causes always determine their effects in one way and no other, and since God determines those causes, nothing that happens falls outside of the necessary determination of causes. There is no question of God supplementing the causal order of the world through special action. Any supplementation would, once again, be self-

325 Spinoza, *Ethics*, IP17C2.
326 Spinoza, *Ethics*, IP33S2.

supplementation. In addition to Schleiermacher's appropriation of Spinoza for positive use, he also adopts his critique of miracles.[327] For both Spinoza and for Schleiermacher after him, a miracle can only be an event, the causes of which we are presently ignorant, but which happens, and is in principle explicable, through the natural order.[328]

The indispensable premise of each of Spinoza's arguments against miracles is A3, that "From a given determinate cause the effect follows necessarily; and conversely if there is no determinate cause, it is impossible for an effect to follow."[329] By categorizing this premise as an axiom, Spinoza implies its truth is undoubted and its application obvious. It is the sort of thing which requires no basis for belief other than adequate comprehension of the terms in use. In consequence, he gives no argument for this axiom, only arguments from it. Of course, that does not mean it is an unsupported claim in the sense of being held only by fiat. It is not a *mere* insistence. Instead, its axiomatic status is a condition for the possibility of the intelligibility of the world. It was an axiom that, in one version or another, both he and many of his theological and philosophical opponents assumed. And, indeed, it is an axiom that anyone wishing to hold that God is the providential creator of the world would not want to do without.

From the axiom that determinate effects follow necessarily from determinate causes, a certain kind of world emerges. Since by A1 and A2, "Whatever is, is either in itself or in another" and "What cannot be conceived through another must be conceived through itself,"[330] all finite effects can only be understood in relation to antecedent finite causes, and so on, and that the whole causal and explanatory chain terminates in God. Since things have determinate natures, and thus determinate powers coordinate with their natures, the world is simply the manifold relation of things and ideas[331] as they bear upon one another in determinate ways.

Once the world is known as a manifold of determinate causes issuing in determinate effects, the pressure on those wishing for miracles as emendations or supplementations of that world begins to build. It is a world where causes and effects are all necessarily determined in a contiguous causal chain. And, if from a given cause *c*, effect *x* necessarily follows, then any alteration of *x* will negate cause *c*. That is, since *x* *necessarily* follows from *c*, in order for *x* *not* to follow,

---

327 *GL* §47.2; *The Christian Faith*, 181–83.
328 Spinoza, *Theological-Political Treatise*, 84; *GL* §47; *The Christian Faith*, 178.
329 Spinoza, *Ethics*, IA3.
330 Spinoza, *Ethics*, IA1, A2.
331 Spinoza, *Ethics*, IIP7.

something (other than *c*, by P27[332]) will have to render *c* undetermined in order to negate *c*'s powers. In short: if *c* then necessarily *x*; if not *x*, then not *c*. The logic is the same if one substitutes systems of causes and systems of effects. So long as all causes and effects are all related and all determinate, the result is the same, since *all* determinate causes will still remain components of the total causal nexus from which determinate effects follow. So that if, for example, a causal system consists in causes *a*, *b*, and *c*, if they all bear upon one another, they will do so in a determinate way and result in a determinate effect or effects. And so, if even only one cause, say cause *b*, were altered, the causal aggregate resulting from *a*, *b*, and *c* would still be different, and so its effects determined otherwise. In short: if *a*, *b*, and *c*, then necessarily *x*; if not *x*, then not *a* and/or *b* and/or *c*. Regardless of how complex the system, if exhaustively causally determined, then any alteration of an effect would negate at least one of its contributing causes. Of course, in and of itself a determinate causal nexus is not an argument against supernatural intervention. One might simply hold that alteration of the causal chain is one of God's prerogatives. It is, after all, God's world.

However, when combined with Spinoza's earlier argument that all that happens follows from the divine essence by natural necessity, the argument becomes devastating. For any alteration of effect *x* will imply, as above, a negation of cause *c*. But cause *c* is itself *necessarily* determined by some antecedent cause, which is itself *necessarily* determined, and so on, all the way back to the divine essence which necessarily determines the whole series. So not only will the alteration of *x* result in the negation of cause *c*, but will also negate *all* antecedent causes prior to *c* which contributed to the whole chain of proximate causes and their effects. Moreover, that whole chain of causes will have issued from God as First Cause. Since, as a consequence of Spinoza's argument above, God is a *determinate* cause which will always issue forth in the entire chain of determinate effects, any alteration of one of the effects in that chain will negate not only its proximate cause, but the totality of its antecedent causal contributors, terminating in God. In other words, cause *c* is not insulated from the divine essence. Instead, it is contiguous with God's causal power. Cause *c* is really c^{x} in some causal chain where God is C^1 followed by the whole causal chain that follows of necessity. So in the final analysis: if C^1, c^2, c^3... c^{x} then necessarily *x*; if not *x*, then not c^{x}... C^1. Any alteration of effect *x* results in a negation of the First Cause, which is the divine essence. Since the divine essence is eternal, immutable, and *a se*, the negation of *x* is absurd. Any miracle would contradict the necessity of the divine essence.

332 Spinoza, *Ethics*, IP27.

Further, Spinoza presses on the advocates of divine supplementation or emendation the equally devastating question of *who* is altering determinate causes to change their effects. It can't be finite agents, for their causal efficacy falls within one and the same order. They are simply one more cause in the causal nexus that comprises the world. Surely it can only be God who can render such effects undetermined, since God alone is not a cause amongst causes, but is absolutely the First Cause of the whole order. But, as above, those effects are already necessary determinations *of God.* Any alteration of effects involves the negation of causes which must ultimately be traced back to one and same cause, which is now the cause of their undoing. Thus, any alteration of the determinate order of things would not only be a contradiction of the divine essence, but a *self*-contradiction. The divine nature would limit itself, and, in so doing, differ from itself. In consequence, God would not be one, because God would now be two; God would not be free, because there would be something over and against the divine nature; and God would not be God because God would not be and be conceived through the divine nature alone. Since the self-negation of God's effects is absurd, God cannot alter the determinate causal series.

However, it might be objected that the above discussion has all been couched as negation. Such language might appear to stack the deck and so beg the question: could not God supplement the causal order without negating it? While perhaps initially promising, this strategy amounts to the same thing. First, it is important to note that even the above argument does not depend on the negation of an effect in the sense of having nothing happen when something would have happened should a miracle not have occurred. Rather, all it requires is an *alteration.* That alteration could indifferently take the form of a supplementation *or* emendation. That is, a miracle could either take the form of something not happening when it would have happened otherwise, *or* something happening when it would have not happened otherwise. Each case applies equally well. This is because what is being negated is the *particular determination* of the effect (and so the cause). The problem is change of any kind, not addition versus subtraction. Even if effects were *added* to the original necessary determinations, they would still change the determination of the effect. The effect would no longer be the one thing it was, but would become something else. It would be differently determined. That something else would *not* follow of necessity from its cause. And because what follows takes the place of what would otherwise have followed, the new effect takes the place of its original cause's determinate effects, and so, in that sense, negates the original cause's powers. In either case, it is the *identity* of the original effect which is negated. If effect x ceases to be x, it necessarily becomes not-x. In this sense, any change is a negation. Since any change is as problematic as any other, the absurdities are unavoidable and Spinoza's argument holds.

God exists and is conceived through nothing but God's own essence. God is uniquely and absolutely *a se*, and so uniquely and absolutely free. Because of the absolute freedom of God, nothing different from the divine essence prompts God to action, either extrinsically or intrinsically; rather, God acts by the necessity of the same essence by which God exists. Because nothing can hinder God, whatever is possible for God becomes actual. And, because everything becomes actual in the only way it can, and in the only order it can, the order of the world issues from the divine essence as determinate effect from determinate cause. Because the world follows as necessary effect from necessary cause, by the absolute necessity by which God can be no other God, the world can be no other way.

Schleiermacher on the Natural Necessity of the World

Just as with Leibniz's appeal to divine wisdom, a nearly exact parallel for Spinoza's argument is found in *The Christian Faith.*[333] However, unlike Leibniz's influence, Spinoza's *lifelong* influence is nearly uncontested.[334] As with Spinoza, Schleiermacher argues from the divine aseity for the principle of plenitude. By reducing the distinction between the possible and the actual in God, he undercuts the hypothetical necessity of created things by making them necessary effects of God. Just as for Spinoza before him, Schleiermacher argues that God could create no other world than this one. God as a determinate cause necessarily renders this world and only this world as a determinate effect.

Like Spinoza, Schleiermacher begins with a consideration of divine freedom. Correlated to our absolute dependence on God is God's absolute independence.[335] That is, God is dependent on nothing, just as all is dependent on God. That aseity *is* God's freedom. Schleiermacher's conclusion, then, is two-fold:

---

333 Examples of previous scholarship which has recognized Schleiermacher's indebtedness to Spinoza on this matter include Lamm, *The Living God*, and Dembski, "Schleiermacher's Metaphysical Critique of Miracles".

334 In contrast, Mariña argues that while the young Schleiermacher was heavily influenced by Spinoza, the elder Schleiermacher was not. While Mariña focuses on Schleiermacher's view of the self, that subject cannot be isolated from his views of God and the world. Given both the clear strain of argument he makes in *The Christian Faith*, as well as a number of quotations which could be martialed in support of his exhaustive determinism, her case seems doubtful, at least when applied to *The Christian Faith*. See Mariña, *Transformation of the Self in the Thought of Friedrich Schleiermacher*, 10.

335 *GL* §§50.3, 54 postscript; *The Christian Faith*, 199, 218–19.

> first, that the entire system of Nature, comprehending all times and spaces, is founded upon the divine causality, which as eternal and omnipresent is in contrast to all finite causality; and second, that the divine causality, as affirmed in our feeling of dependence, is completely presented in the totality of finite being, and consequently, everything for which there is a causality in God happens and becomes real.[336]

In Spinoza's language, nothing can be or be conceived apart from God; and everything which God can conceive, God brings about.

The first idea follows from our absolute dependence upon God. If God plus some other cause was the reason for things being as they are, we would not be absolutely dependent on God, but only relatively so. But since our absolute dependence on God is assumed for the purposes of Christian dogmatics,[337] God must be that on which *all* is absolutely dependent. That dependence requires only God's aseity and God's uniqueness and some notion of causality, however broad, that can account for the dependence of the world. While the notion that nothing falls outside the divine causality itself is not particularly noteworthy, Schleiermacher's use of the notion is: he not only presses the idea to its full extent but also uses it to support the second idea contained within the attribute of omnipotence.

The second idea implied by God's omnipotence, that, "everything for which there is a causality in God happens and becomes real,"[338] is a formulation of the principle of plenitude: that there is no difference between the actual and the possible in God. Rather, God does all that is in God to do. It is Schleiermacher's aim in his discussion to show that this much more controversial principle of plenitude is, as much as the dependence of all on God, entailed by the basic concept of God's independence. If all things are absolutely dependent on God, then God alone is *a se*; but if God is independent, then all that is possible for God "happens and becomes real."[339] There is no realm of the possible outside the actual.[340]

Schleiermacher's rendering of omnipotence into a principle of plenitude is supported by a handful of different but related reductions, including: a reduction of the distinction between the actual and possible in the created order; a reduction of the distinction between God's will and power; and, finally, a reduction of the distinction between the various ways of parsing God's willing and

---

336 *GL* §54; *The Christian Faith*, 211. Translation revised.
337 *GL* §§32, 33; *The Christian Faith*, 131, 133–34.
338 *GL* §54; *The Christian Faith*, 211.
339 "wirklich wird und geschieht." *GL* §54; *The Christian Faith*, 211.
340 See Lamm, *The Living God*, 150.

knowing. All of these distinctions are ultimately unsatisfactory. What remains once all these distinctions have been shown to be untenable is the simple formula that, "everything for which there is a causality in God happens and becomes real."[341] The first idea, the absolute dependence of all on God, and God's correlated independence, is the basis of his reductions which support the principle of plenitude.

Schleiermacher's first attack is on the concept of the potential outside the actual in created things. While he explains that, according to his theology, no demands can be made beyond an account of the natural order,[342] it might nevertheless be objected that "what we call 'all' consists of the actual and the possible [*dem wirklichen und dem möglichen*] and omnipotence must therefore embrace both of these; but that if it presents itself completely and exhaustively in the totality of finite being, then it embraces only the actual and not also the possible."[343] That is, a hypothetical objector might point to a possible problem with Schleiermacher's formulation: if the true *all* is the actual *plus* the possible, then the very notion of an *exhaustive* omnipotence is self-defeating. For if the exhaustive omnipotence of God makes the actual alone to *be*, it would have left out the possible. To our objector this would not amount to an exhaustive omnipotence, but a truncated one. "But," as Schleiermacher replies, "how little the difference between actual and possible can exist for God will appear very clearly, if we only notice in what cases we chiefly apply it ourselves."[344] Schleiermacher gives two different kinds of cases. One kind are those cases where the distinction between the actual and the possible is a distinction between individual and species. The other kind of cases are those where the distinction is between that which is possible according to a thing's nature but which never develops because of outside hindrances.[345]

The former cases of distinguishing the actual and the possible in species are based on an abstraction. Of any given species, its members will be variously determined—they will have certain characteristics and not others. For example, a person might have two entirely brown eyes or two entirely green eyes, but both of the same person's eyes will not be entirely brown *and* entirely green at the same time. However, we still call both brown and green eyes *potential* in regards to the species, even though in the case of any given individual those determinate characteristics will be mutually exclusive, since humans *as*

341 *GL* §54; *The Christian Faith*, 211.
342 *GL* §54.2; *The Christian Faith*, 212–13.
343 *GL* §54.2; *The Christian Faith*, 213. Translation revised.
344 *GL* §54.2; *The Christian Faith*, 213. Translation revised.
345 *GL* §54.2; *The Christian Faith*, 213.

a species do in fact have both brown and green eyes. But for God, possessed of exhaustive foreknowledge,[346] "the species exists originally as the sum-total of its individual existences, and these in turn are given and established together with their place [*Ort*] in the species, so that what does not hereby become actual, is also, so far as He is concerned, not possible."[347] The ordinary use of the notion of the "possible" in this case results solely from ignorance, from the fact that we don't know of every particular determination of any given species, and so must make a generalization on the basis of our merely partial knowledge. However, God is not subject to ignorance. Since the ignorance that warrants the abstraction is inapplicable to God, the abstraction itself is also inapplicable.

The second kind of case is similar. However, instead of resulting from a generalization of a species from its members, this case results from the generalization of the native powers of an individual. "In the same way," says Schleiermacher, "we say that much is possible by virtue of the nature of a thing (when we take together its determinations by its species and as an individual being), which yet does not become actual because it is hindered by the position of the thing in the sphere of general interaction."[348] For example, we might say a person could *possibly* have been a football star, but, in fact, they never become one because of a career-ending injury. The possible, in this case, is an abstraction from the concrete circumstances that actually occur. And as useful as this way of thinking is in considering finite creatures, it cannot apply to God since:

> [I]f we could have taken into account for each point the influence of the whole system of interaction, we should then have had to say that what was not actual was also not possible within the system of nature. In God, however, the one is not separated from the other, that which exists for itself having one ground and the system of interaction another, but both these are grounded with and through each other, so that in relation to Him only that is possible which has its foundation in both equally.[349]

It is again human ignorance alone which considers an individual without exhaustively considering its circumstances—an ignorance which in no way applies to God. Tellingly, both reductions are, upon inspection, applications of conceptual containment. Conceptual containment is supported in each case by appeals to exhaustive divine foreknowledge which itself follows from the divine aseity.

346 *GL* §55.1; *The Christian Faith*, 219–22.
347 *GL* §54.2; *The Christian Faith*, 213. Translation revised.
348 *GL* §54.2; *The Christian Faith*, 213.
349 "[...] *so daß in Beziehung auf ihn nur dasjenige möglich ist, was in dem einen von beiden eben so sehr begründet ist wie in dem andern.*" *GL* §54.2; *The Christian Faith*, 213.

Nevertheless, thinkers like Leibniz held that, once chosen, everything that happened did so in one way and no other, and yet were able to support a robust notion of hypothetical possibilities. Schleiermacher's criticism has so far left such positions unscathed. But possibilities of this sort have their own vulnerability: namely, that an idea of the possible beyond the actual relies on God's intellect extending beyond his will. It is that question which Schleiermacher takes up next.

Schleiermacher's first move is to rule out inadequate notions of intellect and will in God. "[O]ur first rule," he says,

> must be to exclude from the spirituality of the Divine Essence everything which necessarily contains in itself receptivity or passivity. Therefore, just as the divine will must not be thought of as a faculty of desire, so the divine omniscience must not be considered as a perceiving or experiencing, a thinking together or a viewing together [*ein Vernehmen oder Erfahren, ein Zusammendenken oder Zusammenschauen*].[350]

These negations of imperfection again follow from God's aseity. If God's will were a faculty of *desire*, God would necessarily act for some end which God lacked, a lack which would condition God—the very dilemma Spinoza sought to press upon advocates of divine willing.[351] And, if God's intellect were a passive perceiving or experiencing, then such knowledge would compound to the divine essence. In both cases, God would be dependent on something else.

Not only must the compounding of extrinsic objects of knowledge and desire be excluded from consideration of the divine attributes, but also the compounding of the attributes themselves. Instead, God is absolutely simple.[352] The distinct attributes are simply one and the same essence first considered one way, then another.[353] The divine essence is itself completely undivided. "Hence the divine thinking is the same as the divine will, and omnipotence and omniscience are one and the same."[354] Any attempt to introduce a real difference between attributes once again contradicts the aseity of God. If attribute was different from attribute, or attributes different from the divine essence, something other than Godself would determine God's knowledge, or God's willing, or both. In order to avoid this dilemma, Schleiermacher follows the classical tradition and sees the attributes in strict identity with the divine essence, and so with one another. The result is simple: "God knows all that is; and all that God knows is, and these

350 *GL* §55.1; *The Christian Faith*, 220.
351 Spinoza, *Ethics*, IApp.
352 *GL* §56; *The Christian Faith*, 231.
353 *GL* §50; *The Christian Faith*, 194–200.
354 *GL* §55.1; *The Christian Faith*, 221.

two are not two-fold but single; for His knowledge and His almighty will are one and the same."[355] Any attempt to see possibilities in the divine knowledge which are not in the divine will runs afoul of the identity of attributes.

Any remaining attempt to secure possibilities in God must rely on a distinction between God's will and God's power. As with Spinoza before him, Schleiermacher argues such a distinction again contradicts the divine aseity and so is absurd. On the one hand, the identity of attributes rules out such possibilities from the start—for if God's power, like all the divine attributes, simply *is* the divine essence, there can be no difference between God's will and God's power. On the other hand, any attempt to avoid such identity results in an introduction of something different from the divine nature which conditions it. And so, "There is, however, as little distinction between 'can' and 'will' in God, as between the actual and the possible. For whichever is greater than the other, the will or the ability, there is always a limitation, which can only be done away with if both are made equal in range."[356] Since, by the divine aseity, nothing determines God to act but God's essence, God's power and will must be coextensive. In consequence, everything that God wills comes to pass. And since *everything* God wills comes to pass, what does not come to pass, God does not will. God knows and wills and acts by one and the same necessary essence by which God exists.

And yet, once again, such necessity might seem opposed to divine freedom. Schleiermacher agrees that such worries are warranted, but only, "when an active and an inactive, and free and a necessary will are set over against the other."[357] In that case, "The *necessary* will would be related to what God wills in virtue of His essence, the *free* to that which, so far as his essence is concerned, He could just as well not will; where it is assumed that it does not belong to His essence to reveal Himself."[358] If such contrast were assumed, if God necessarily wills what God wills, it seems God does not do so freely.

But again, just like Spinoza, Schleiermacher argues that such a distinction is inadmissible:

> [I]t seems that this contrast [between the necessary and the free] cannot be applied to God at all, and what has been brought under the contrasted heads is not really separable. For where such a contrast exists the necessary must be unfree, and the free be grounded in no necessity, and so arbitrary. Each, however, is an imperfection; and consequently this con-

---

355 *GL* §55.1; *The Christian Faith*, 222.
356 *GL* §54.3; *The Christian Faith*, 214.
357 *GL* §54.4; *The Christian Faith*, 216.
358 *GL* §54.4; *The Christian Faith*, 216–217.

> trast has its place solely in that existence in which each being is co-determined by the rest. We must therefore think of nothing in God as necessary without at the same time positing it as free, nor as free unless at the same time it is necessary.[359]

Since necessity and freedom are not contraries in God, by absolute freedom God wills what God wills (which is what God knows) of absolute necessity.

With this reduction, Schleiermacher explicitly intends to deny hypothetical necessity. "Since," Schleiermacher says,

> in relation to God no distinction between the potential and the actual can be allowed, it is easy to pass judgment on the popular expression of God's omnipotence, which has often been adopted even in scientific discussions, namely, that it is the attribute in virtue of which God is able to effect all that is possible, or all which contains no contradiction in itself.[360]

Notice that the definition of the possible here is the *exact* definition given by Leibniz and others in the hypothetical necessity tradition. Schleiermacher's response is a plain rejection. "If, of course," he says, "contradiction is taken *realiter* and that is called contradictory which can find no place in the whole of existence, this is perfectly correct; for all the compossible is certainly produced by the divine omnipotence."[361] What Schleiermacher seems to grant in this definition of the possible, he does only on Spinoza's grounds. And so what initially appears to be a partial concession actually turns out to be yet another reiteration of the principle of plenitude.

The will of God, which must come to complete expression, is a determinate will, expressing itself in determinate effects. Like Spinoza, Schleiermacher holds that those effects themselves are determinate causes likewise issuing in determinate effects. And so, the natural order is one mutually determined whole, *absolutely* willed by God as a whole. Here Schleiermacher deserves quoting at length:

> But it is by no means the case that God wills some things absolutely, others conditionally; just as with regard to every event there is something of which one can say, if this were not that event would not be; so with regard to every individual thing—the fact that it exists and it exists in this way—we can say that God wills it conditionally, because everything is conditioned by something else. But that whereby something else is conditioned is itself conditioned by the divine will; indeed in such a way that the divine will upon which the conditioning rests, and the divine will upon which the conditioned rests is not different in each case, but one only and the same; it is the divine will embracing the whole framework of

359 *GL* §54.4; *The Christian Faith*, 217.
360 *GL* §54.3; *The Christian Faith*, 214.
361 *GL* §54.3; *The Christian Faith*, 214.

> mutually conditioning finite being: and this naturally is the absolute will, because nothing conditions it. In this way everything individual would be willed by God conditionally, but the whole willed absolutely as a unity. On the other hand, if for once we take an individual out of the order, and relate it *so* to the divine will, we shall have to say that each individual existing for itself, so far as we regard it not as conditioned by, but as co-conditioning the whole, is so fully willed by God as what it is, that everything else must be so, and cannot be otherwise than as follows from its action; which is as much as to say that it is absolutely willed by God. In this respect, therefore, it can be said that every individual, so far as it must be affected by the rest, is also only conditionally willed by God; but, of course, not as though on that account it were any less willed, or any the less came to reality. Everything, however, so far as it is itself effective, and in various ways conditions other things, is absolutely willed by God.[362]

In light of the mutual determination of finite things in relation to one another, such that everything must be as it is in order for everything else to be what it is, and in light of the necessary determination of this whole by God, all notions of the merely possible disappear. For, "that, the existence of which conflicts with the existence of all else, is also contradictory to itself. Thus there is no divine knowledge of it even on the traditional explanation of the divine omniscience, for the self-contradictory is neither a thing nor cognizable."[363] When completely described, the very notion of the possible outside the actual rules itself out from the start. God has no knowledge of nothingness. And, "As soon, then, as we express it so, namely that God knows what would have resulted if at any point the impossible had become real, this knowledge, as a whole, dissolves into nothing, because what rests solely upon the becoming real of the impossible is itself impossible."[364]

Instead, again following Spinoza, Schleiermacher holds that God acts by the necessity of the divine nature alone. Since freedom and necessity are one in God:

> Just as little, however, can we think of God's willing Himself, and God's willing the world, as separated the one from the other. For if He wills Himself, He wills himself as Creator and Sustainer, so that in willing Himself the world is already included; and if He wills the world, in it He wills his eternal and ever-present omnipotence, wherein willing Himself is included; that is to say, the necessary will is included in the free, and the free in the necessary.[365]

362 *GL* §54.4; *The Christian Faith*, 216.
363 *GL* §55.2; *The Christian Faith*, 224.
364 *GL* §55.2; *The Christian Faith*, 225.
365 *GL* §54.4; *The Christian Faith*, 217.

As God acts by the necessity of his nature, the world follows of absolute necessity as determinate effect from determinate cause.

If the world follows as an integral whole as effect from God as cause, then Schleiermacher's adoption of Spinoza's critique of miracles is easy to follow. It proceeds in the exact same way.[366] Since any alteration of the determinate course of nature negates its antecedent causes, and since those causes are all linked in one contiguous chain back into the divine essence (as above), any alteration of the course of nature contradicts the necessity of the divine essence. And since only God could alter the course of things in such a way, any such contradiction of the determinate order of causes would be a self-contradiction. Instead, the world follows its ordained, immutable, and indeed, necessary, course. Schleiermacher adopts Spinoza's argument against miracles alongside Leibniz's and deploys both in mutual support.

Spinoza's and Schleiermacher's Ends

A central aim of Spinoza's *Ethics* is the reduction of hypothetical necessity to natural necessity. Likewise, Schleiermacher spends a large portion of *The Christian Faith* making the same case. The motivation behind Spinoza's effort was, however, more obvious: Spinoza not only sought to exclude miracles, but also sought the more general dismissal of final causes from divine action—particularly the crudely anthropocentric forms of final causes with which he was familiar. As Schleiermacher was also motivated to challenge inadequate views of divine action (including miracles), there is at least some common ground. And yet, as we saw last chapter, Schleiermacher is no enemy of final causes. In fact, he argues that without a determinate *telos* God's action is unintelligible. It seems that Schleiermacher's and Spinoza's respective motivations are incompatible. But that would make Schleiermacher's use of Spinoza's argument, that everything happens by natural necessity, very strange indeed, since it appears Spinoza sought that very reduction *in order* to dismiss final causes in divine action altogether.[367]

---

366 *GL* §47.2; Dembski rightly notes the detailed parallels of Spinoza and Schleiermacher's respective arguments and how each is also supported by a very specific (and nearly identical) doctrine of God. However, Dembski's conclusion that Spinoza and Schleiermacher beg the question is only true if one artificially abstracts their arguments against miracles from their reasons for their particular doctrine of God. See Dembski, "Schleiermacher's Metaphysical Critique of Miracles," 463–464.

367 Spinoza, *Ethics*, IApp.

The solution to this puzzle will be taken up in more detail in the following chapter. There I argue that the eternal covenant is incomplete without the unity of both teleologically determined action, and metaphysically necessary action. But, for now, it is worth noting exactly what Spinoza thought he was dismissing and how it relates to varying views of final causes.

The heart of Spinoza's attack on final causes in the *Ethics* is IP33S2, where Spinoza tells us his position eliminates the freedom some assign to God—"namely, an absolute will."[368] It was this will which was subjected to reduction above through consideration of the divine aseity and uniqueness in combination with the identity of the divine attributes with the divine essence. For, in order for God's will not to *be* God's essence, it would have to differ and so limit that essence, which is absurd. Such a conclusion contradicts the divine aseity and uniqueness. But, if not different, the divine will *is* the divine essence; if the divine essence, it necessarily is such as it is. In order for God's will to be other than it is, God would have to be another God, which is also absurd. So, Spinoza, concludes, God cannot will for an end. But notice: in this analysis Spinoza *assumes* that if God acts for the sake of an end, that end would be distinct from God's self. But the end for which God acts need not be "something outside God, which does not depend on God, to which God attends, as a model, in what he does [...]."[369] Rather, the end for which God acts *is* God's very essence—as the tradition had long insisted.[370] Therefore, on both counts, Schleiermacher agrees that God cannot act for the sake of an *extrinsic* end; but since God does act for the sake of an end, that end must *be* the divine essence. Anything else would contradict the whole strain of his argument. Since everything depends on God, and God on nothing, God does not act for an end which is other than the divine essence. Spinoza and Schleiermacher both attack a deficient account of final causes. But that view of final causes is not the only option, and, instead of rejecting teleological divine action altogether, Schleiermacher takes a third way.

Nevertheless, Spinoza's reduction of the distinction between absolute and hypothetical necessity was a response to a genuine problem. And, though related to the question of final causes, it is distinct from it—as shown by Schleiermacher's simultaneous adherence to teleological divine action *and* Spinoza's reduction. For Leibniz, hypothetical necessity was supported by genuine possibilities. Even if God could choose no other, God still chooses from amongst possibles. The effect of this distancing between the divine essence and the divine

368 Spinoza, *Ethics*, IP33S2.
369 Spinoza, *Ethics*, IP33S2.
370 See Thomas Aquinas, *Summa Theologiae*, Ia, q. 19, a. 2, ro. 2.

decree was to free God from the strictures of conceptual containment, and instead to secure God transworld identity.[371] At least hypothetically, God could be the same God even if the world were other than it is. It was *this* view which Schleiermacher most eagerly sought to eliminate, and he follows Spinoza in the reduction of hypothetical necessity to natural necessity *precisely to deny God transworld identity* by denying that there are any possible worlds for God to have identity in. In the following chapter, we will see what extraordinary consequences Schleiermacher sought to secure by this denial, and how his combination of a teleologically ordered world and the natural necessary of this world secures the benefits of the eternal covenant.

371 See Robert Merrihew Adams, *Leibniz: Determinist, Idealist, Theist* (Oxford: Oxford University Press, 1994), 55.

Chapter 6 The Self Presentation of the Divine Essence: Schleiermacher on the World as the Artwork of God

"We may now say that our discourse about the nature of the universe has an end. The world has received animals, mortal and immortal, and is fulfilled with them, and has become a visible animal containing the invisible—the sensible God who is the image of the intellectual, the greatest, best, fairest, most perfect—the one only-begotten heaven."

- Plato, *Timaeus*

When he laid out his position on miracles in his *Theological-Political Treatise*, Spinoza did not only make a negative case. True, he held, "that nothing happens contrary to nature, but nature maintains an eternal, fixed immutable order,"[372] and so miracles, as the tradition understood them, were impossible. But this was only half the story.

The other half was what the inviolable natural order offered in knowledge of God. Spinoza explained, "that from miracles we cannot know about either the essence or the existence or the providence of God, but rather all three are much better grasped from the fixed and unchangeable order of nature."[373] His perhaps initially puzzling claim was fleshed out as follows:

> [S]ince we know that all things are determined and ordained by God, and that the operations of nature follow from the essence of God, and the laws of nature are the eternal laws and volitions of God, we must conclude, unconditionally, that we get a fuller knowledge of God and God's will as we acquire a fuller knowledge of natural things and more clearly understand how they depend on their first cause and how they behave according to the eternal laws of nature. From the perspective of our understanding, hence, we have much more right to term those phenomena which we understand clearly and distinctly works of God and attribute to them the will of God, than works of which we are wholly ignorant, however strongly they grip the imagination and make us marvel.[374]

---

372 Spinoza, *Theological-Political Treatise*, ed. Jonathan Israel (Cambridge: Cambridge University Press, 2007), 82.
373 Spinoza, *Theological-Political Treatise*, 82.
374 Spinoza, *Theological-Political Treatise*, 85–86.

DOI 10.1515/9783110542301-006

It is investigation of the natural order, and comprehension of the world under that order's laws which, more than miracles, grants knowledge of God. The reverent ought look to the rule, not the exception.

Spinoza's argument is clear enough. If the world follows a determinate causal order all the way back into the divine essence, then miracles are incoherent. But, by the same necessity by which miracles are rendered absurd, effects follow from causes deductively. Given the divine will, the natural order necessarily follows. That deductive relation enables an inference from the contents of the world to the divine will. And as the divine will likewise follows necessarily from the divine essence, the natural order follows necessarily from the divine essence. Since the world follows necessarily from the divine essence, the order of causes is reversible back into the divine essence. Because the natural order follows of necessity in the way it does and no other, the more that order is understood, the more the divine essence is known.

If, and only if, the world follows necessarily from the divine essence, is it possible to make this inference. Anything less fails to provide a sufficient logical basis. At the same time, it is this very necessity which supports Spinoza's critique of miracles. The two go hand in hand. Without the one, the other cannot be had. Accepting what the tradition would see as a burden in denying miracles is not done solely for the sake of criticism or novelty, but rather is necessary to accrue substantial theological benefits.

The great gains of such an approach suggest a motivation for Spinoza's argument against miracles beyond mere criticism. It is not obvious, anyway, that he first and foremost sought to curtail (as much of the tradition would see it) divine action, and then only later by chance happened upon a window into the divine essence. A more likely story is that just as the logic of both positions meets at the same point, so too does their motivation. In both Jewish and Christian traditions, piety has often motivated iconoclasm. Since Schleiermacher, once again, makes the same argument in *The Christian Faith* and, importantly, does so for explicitly pious reasons, his arguments against miracles, for the necessity of the world, and finally for the world as the revelation of God, are connected in the same way. But a quandary remains: Spinoza understood his position to eliminate teleological divine action, while Schleiermacher not only accepts final causes but argues that divine action is unintelligible without them. In adopting Spinoza's arguments for absolute necessity, Schleiermacher is then obliged to explain how this necessity and his subscription to final causes are coherent.

The aims of this chapter are threefold: one, to explain how Schleiermacher unites Leibniz's insistence on final causes with Spinoza's denial of hypothetical necessity (the arguments of Chapter 3 and 5, respectively) in spite of their *prima facie* incompatibility; two, show how all steps in Schleiermacher's account, which entails the denial of absolute miracles, are not only motivated by criticism but by great theological gain; and three, to entertain the objection from the problem of evil that this account naturally raises. With only minor adjustments or clarifications Schleiermacher combines two powerful, seemingly exclusive, arguments against absolute miracles into a single compound case while, at the same time, the unity of teleology and necessity yields his notion of the world as the artwork of God, the perfect work of the perfect artist, and so the absolute revelation of God.

The Unity of Teleology and Necessity

Not all accounts of end-ordered action are alike, and few cohere with the absolute divine self-subsistence and simplicity to which Schleiermacher is committed. Thus, Schleiermacher's joins Spinoza in rejecting a certain sort of teleological divine action. If the supposed end for which God acts is an end of need, something which God lacks and must acquire, or if that end and the means to it are accidental to God, grafting that action for an end through those means onto the divine essence, God cannot act for the sake of an end. In order for Schleiermacher to coherently maintain his claim that divine action is, in fact, teleological, he has to give an account that does not imply the dependence of God on an extrinsic good.

The first step in Schleiermacher's account is to specify the end for which God acts. In so doing, he also specifies what sort of end it is. Schleiermacher holds that the *telos* of divine action is love, that is, the divine self-impartation. "The divine love," Schleiermacher says, "as the attribute *in virtue of which the divine nature imparts itself*, is seen in the work of redemption."[375] And, he further explains:

> For love is but the impulse [*Richtung*] to unite self with others and to will to be in others; if then the pivot of the divine government is redemption and the foundation of the Kingdom of God, involving the union of the Divine Essence with human nature, this means that the underlying disposition can be presented only as love.[376]

375 "*Die göttliche Liebe als die Eigenschaft, vermöge deren das göttliche Wesen sich mittheilt, wird in dem Werk der Erlösung erkannt.*" *GL* §166; *The Christian Faith*, 727. Emphasis added.
376 *GL* §165; *The Christian Faith*, 726. Translation revised.

This exactly follows the traditional teaching that the creation and redemption of the world were for the sake of God's self-communication.[377] According to this traditional account, divine action is teleological, and all teleological action is ordered to *acquiring* some good, and so regards a lack. But the lack in the case of divine action is not God's lack, but creatures'. It is we alone who are ordered to acquiring communion with God. God lacks nothing and so cannot act to acquire anything. Instead, the divine essence communicates itself to others out of no need of its own, but out of its plenitude.[378] In this sense, the divine goodness is the end of all things.[379] In making the union of the divine essence with human nature through the impartation of that essence to creatures the *telos* of divine action, Schleiermacher makes plain that the end of divine action is, just as the tradition would have it, ordered to the creatures' perfection, not God's.

Importantly, this sidesteps one of Spinoza's main criticisms. Since the end of divine action is not a *divine* lack, no need compels God, and God is not determined to act by something other than the divine nature itself. Since it is, for Spinoza, the implication of divine lack that makes divine teleological action incoherent, once the threat of this supposed divine lack is dispelled, so is Spinoza's main concern. And yet, although the divine end of self-communication is not an end of need, Spinoza might still worry that this end is, nevertheless, extrinsically motivated, and that, if so, God does not act by the necessity of the divine nature, and so not in perfect freedom. Spinoza does, after all, note that at least some theologians have recognized this problem before and so have insisted that the divine end is not an "end of need" but an "end of assimilation."[380] Otherwise, the divine essence would be prompted to action by something not itself.

Schleiermacher shares this concern, and addresses it by arguing that the divine love is strictly equivalent with the divine essence. "[L]ove alone," Schleiermacher says, "is equated with the being or essence of God. Hence it is in this exclusive form that our paragraph has to be established and justified, namely, that

---

377 "I answer that it is goodness which prompts God to create with the purpose of communicating himself; and this same goodness, when combined with wisdom, prompts him to create the best: a best that includes the whole sequence, the effect and the process." Leibniz, *Theodicy* §228; 269. See also Aquinas, *Summa Theologiae* , Ia, q.19, a. 2, resp.

378 "Further, he [God] acts to do good, and not to receive it. *Melius est dare quam accipere*; his bliss is ever perfect and can receive no increase, either from within or from without." Leibniz, *Theodicy* §217; 264.

379 Aquinas, *Summa Theologiae*, Ia, q.19, a. 3, resp.

380 Spinoza, *Ethics,* IApp.

only love and not another divine attribute can be equated with God."[381] God is essentially love. It is simply the divine nature to be and do so. As such, love stands alone amongst the divine attributes as synonymous with the name "God." "Love and wisdom alone, then, can claim to be not mere attributes but also expressions of the very essence of God [...]."[382] And so God does, after all, act by the necessity of the divine nature alone *precisely* in acting for the *telos* of self-communication. Since that nature turns out to *be* love, the divine essence imparts itself of absolute necessity. God is love, and its opposite implies a contradiction.

While other theologians would agree that the divine goodness is the absolutely necessary ultimate end for which God acts, what is less traditional is Schleiermacher's denial of the merely hypothetical necessity of the *proximate* divine ends and their means.[383] Those proximate ends—fellowship with Christ through his implanted God-consciousness, and that as spread through the church, etc.—serve as means for God's ultimate end of self-communication and impartation. Again (see Chapter 3), as Schleiermacher says:

> Hence we can say, regarding two points formerly made out—both the essence of things in their relations to each other and the order of reciprocal interaction between them—that they exist through God as they exist with regard to the redeeming revelation of God in Christ, by which the spirit is developed to perfection. Everything in our world, namely, human nature in the first place and all other things in direct proportion to the closeness of their connection with it, would have been arranged differently, and the entire course of human and natural events, therefore, would have been different, if the union of the Divine Essence with human nature in the Person of Christ, and, as a result thereof, the union of the Divine Essence with the fellowship of believers through the Holy Spirit, had not been the divine decree.[384]

Importantly, Schleiermacher argues that these *proximate* ends are as absolutely necessary as their distant end. And, equally importantly, those proximate ends are tied, through conceptual containment, to the world as a whole. Again (see Chapter 5), Schleiermacher says:

> [J]ust as with regard every event there is something of which one can say, if this were not that event would not be; so with regard to every individual thing—the fact that it exists and it exists in this way—we can say that God wills it conditionally, because everything is conditioned by something else. But that whereby something else is conditioned is itself condi-

381 *GL* §167; *The Christian Faith*, 730. Translation revised.
382 *GL* §167.2; *The Christian Faith*, 731–32.
383 Cf: Aquinas, *Summa Theologiae*, Ia, q.19, a. 3, resp.
384 *GL* §164.2; *The Christian Faith*, 724. Translation revised.

> tioned by the divine will; indeed in such a way that the divine will upon which the conditioning rests, and the divine will upon which the conditioned rests is not different in each case, but one only and the same; it is the divine will embracing the whole framework of mutually conditioning finite being: and this naturally is the absolute will, because nothing conditions it. In this way everything individual would be willed by God conditionally, but the whole willed absolutely as a unity.[385]

Since Christ and the church only exist in this world, the world as it actually is, is necessary; and, since everything in the world is so mutually determined that to change one thing would change everything else, not only Christ and the church, but *this world* in its every detail is as absolutely necessary as the distant end of the divine goodness. The whole chain of means and ends, distant and proximate, is absolutely necessary to the divine essence.

God's perfection determines distant and proximate ends, and their means. As the denial of divine perfection contradicts the divine essence, so does the denial of the world's perfection as determinate effect issuing from determinate cause. By combining Leibniz's argument for the necessity of the best from God's perfect wisdom with Spinoza's denial of God's transworld identity, Schleiermacher concludes that the perfect wisdom of God results in one world: this world. But, as the perfect ordering of God's works, it is, in fact, *not* "best" because it could not be otherwise:

> [W]e must stop at the affirmation that the world is *good*, and can make no use of the formula that it is *best*; and this because the former assertion signifies far more than the latter. The latter expression is connected with the idea (which we have already rejected) of many worlds all originally equally possible with the one which actually came into existence, and also seeks to represent the entire course of time in the actual world as the result of mediate divine knowledge (the idea of which we have also rejected), so that the whole productive activity of God is assumed to be selective and therefore secondary.[386]

All superlatives are comparatives, and there are no other possible worlds to compare. Any means less than perfect ones contradict the divine essence, and so any world but this one is itself a contradiction of the divine essence. Though the world is the perfectly wise means to perfectly good ends, no world but this one is possible precisely because anything else would contradict the divine goodness.

However, while this might satisfy Spinoza's concerns, it might not yet satisfy Leibniz's. For his part, Leibniz seemed convinced in his correspondence with

---

385 *GL* §54.4; *The Christian Faith*, 216.
386 *GL* §59 postscript; *The Christian Faith*, 241.

Clarke that any action that was *absolutely* necessary was, by definition, not teleological—hence his maintaining the world's merely hypothetical necessity.[387] And, of course, merely hypothetical necessity is precisely what Schleiermacher denies. At the same time, however, Schleiermacher agrees that divine action is most certainly teleological. And, equally importantly, he not only concedes, but insists upon the determinate particularity of that action. Schleiermacher and Leibniz concur on this point. It is, however, for Schleiermacher the determination of action, the assigning of a concrete end and means ordered to that end, *not* the denial of absolute necessity which is the sole necessary condition of a teleological account. In support of this reduction he provides no argument. As his case is most economical it is his opponents' burden to argue for additional notions of freedom necessary for a teleological account—notions which Schleiermacher has shown are not only unnecessary, but burdensome.

As we saw in Chapter 3, there are imperfections implied in ordinary means-ends ordering, imperfections which surely do not apply to divine action, and which Schleiermacher intends to excise from Leibniz's account. First, there is the deliberation between means, which implies ignorance on the part of the agent. Schleiermacher denies this deliberation, along with its sense of the optional, in the strongest possible terms. But there is a second imperfection implied in ordinary means-ends ordering of action: the *use* of means. That is, ordinary means are extrinsic to an agent. They supplement action and increase an agent's powers. In the case of God, both implications are imperfections. "Means," says Schleiermacher, "are never employed except where the agent has to have recourse to something not originated [*nicht hervorgebrachtes*] by himself. Nor can we easily conceive the determination of means otherwise than in the form of choice, which means reverting to that very mediate knowledge which we discarded."[388] Schleiermacher finds any *division* of means and ends, in regard to divine action, problematic.[389] The anthropomorphic notion of God deliberating over means won't do. Neither will a notion of God making use of something ex-

---

387 As Garber points out, "Leibniz at least entertains the position that God might have *chosen* the best of all possible worlds *necessarily* [...]." And yet he takes care to note that, at the same time, Leibniz maintains that this necessity is *not* an absolute necessity precisely to maintain final causes and divine choice. Although Griffin sees hypothetical necessity as having identical logical force as absolute necessity, he too agrees that the maintenance of final causes motivates Leibniz's distinction. Garber, *Leibniz: Body, Substance, Monad*, 236–237, 336. Griffin, *Leibniz, God and Necessity*, 58–59, 65.

388 *GL* §168.1; *The Christian Faith*, 734.

389 *GL* §168.1; *The Christian Faith*, 732–34.

ternal to the divine essence. Both threaten to violate the self-sufficiency which Schleiermacher and Spinoza insist upon. Both threaten to subject God to fate.

And yet, in spite of the imperfection which *ordinary* means-ends language implies, such language remains indispensable. Schleiermacher himself is unable to avoid its use, and, in fact, insists on it. "The divine wisdom," he says, "is the principle which orders and determines the world for the divine self-imparting which is evinced in redemption."[390] The world, according to him, is ordered, and ordered teleologically. Any teleological ordering will have to be ordered means to end: that *by* which a thing is ordered is that *for* which it is ordered. To *wholly* deny the latter would contradict the former. Instead of outright denial, Schleiermacher is, and indeed must be (on pain of self-contradiction), committed to the responsible qualification of those terms in their usual sense. The result is his notion of the world as a divine work of art.

Schleiermacher offers the artist's inspiration in place of the hypothetical, with its roots in deliberation, which he has so forcefully rejected. "Hence," Schleiermacher says, "it would have been far safer, if one does not start from what is human, to transfer to God, illimited and perfect, the certainty of the perfect artist, who in a state of inspired discovery thinks of nothing else, to whom nothing else offers itself, save what he actually produces."[391] In this one sentence, Schleiermacher includes both absolute necessity and complete plenitude. The perfect artist has a total certainty of purpose. The artist's purpose cannot be other. He thinks of nothing but what he actually produces. And what is actually produced is what offers itself to him—no more, and no less. All that he can do, he does. And yet, the necessity of the work is not confused with its dependence. The art is an effect, the artist its cause.

At the same time, the analogy with an inspired artist answers the question of motive—motive without which divine action is indeterminate. For the artist is not prompted to action by an extrinsic cause, but an intrinsic one. The artist produces what he does because he apprehends his work in himself and only then looks to make it real extrinsically. It is his self-apprehension which suffices to motivate, and it does so because what is apprehended is in itself good and beautiful. And because in the case of God—the perfect artist—nothing can be compounded to the divine essence, this intrinsic good cannot be anything but essential. While the human artist might hope for inspiration from without to become his own, God does not begin to be or do anything, and so God does not begin to apprehend the good for which to act. Instead, by nature, God contemplates the su-

390 *GL* §168; *The Christian Faith*, 732.
391 *GL* §55.2; *The Christian Faith*, 225.

preme goodness of the divine essence eternally. It is God's natural, reflexive, self-apprehension which, like the artist's inspired vision, serves as spontaneous prompt to action.

The perfect artist's self-expression is a form of communication. Communication in this case involves both knowledge and union.[392] The artist not only acts to make public his "inspired discovery," to share his vision with others so that by knowing that good they may likewise enjoy it, but actually shares himself for others' sakes. The good apprehended and expressed is the artist's own. By expressing his inspiration in his art, the artist expresses himself. Thus, when another comes to know the art, they come to know the artist. Artistic expression is, therefore, self-communication in the strongest sense: what the artist communicates *is himself.*

In the case of even the best human artist, however, this self-communication has its imperfections. The artist only ever shares himself incompletely, and the inspiration that the artist communicates is itself only discovered *by the artist* over time. God's artistry, however, suffers from no such imperfections. God's knowledge is not discursive. By the divine omniscience all that can be known, is so.[393] And by the divine omnipotence all to which God's power extends, "happens and is made real."[394] In consequence, the divine artwork communicates the divine nature completely. The divine artwork, whatever else it may be, is the unhindered, complete expression of the divine nature, which is expressed of natural necessity.

The divine self-communication is the divine love. Again, as Schleiermacher says, "The divine love, as the attribute in virtue of which *the divine nature imparts itself*, is seen in the work of redemption."[395] It is the natural necessity of this love that authorizes the unique equation of the divine love with the divine essence.[396] But, in isolation from the divine wisdom, love is an abstraction of the end of divine action from its means. Instead, Schleiermacher says, "The divine wisdom is the principle which orders and determines the world for the divine self-imparting which is evinced in the work of redemption."[397] And, "From the fact that we take the divine love as being also wisdom it follows, first of all, that we cannot possibly regard all finite being in its relation to our God-con-

392 Hence Schleiermacher's description of this communication as both "presentation" and "impartation." *GL* §168; *The Christian Faith*, 732–33.
393 *GL* §55.1; *The Christian Faith*, 222.
394 *GL* §54; *The Christian Faith*, 211.
395 *GL* §166; *The Christian Faith*, 727. Emphasis added.
396 *GL* §167; *The Christian Faith*, 730.
397 *GL* §168; *The Christian Faith*, 732.

sciousness except as (whatever else may be the meaning we give to the term 'world') an absolutely harmonious divine work of art."[398] Love is the world's end, and wisdom its order. Only in their unity do the two comprise a work of art. And it is only as an *artwork*, as a unity of means and ends, that the world is a self-communication of God.

In spite of Spinoza's disavowal of divine teleological action, few have noted how alike Schleiermacher's doctrine of the divine self-communication is to Spinoza's treatment in the *Ethics*. Lamm takes up the comparison, but her conclusions on this particular point are mistaken. She claims that Schleiermacher "makes a clear departure from Spinoza."[399] The substance of this departure is that "Spinoza refuses to apply the attribute of love to God,"[400] whereas Schleiermacher makes love essential. Spinoza does, of course, deny the *affect* of love to God (since, by virtue of God's perfection, God can neither suffer nor enjoy change). However, Spinoza nevertheless does attribute love to God when he claims that *God loves himself with an infinite intellectual love,* and that the love by which God loves himself is *the love by which God loves us.*[401] And, since God does all that God does by natural necessity, the intellectual love of the divine nature and of us in that nature is also, therefore, essential. Moreover, these reflexive layers of divine love are warranted by causal relations and explained in causal terms, further contradicting Lamm's claim that Spinoza does not proceed by the *via causalitatis*. The effect of this misreading is both to artificially distance Schleiermacher from Spinoza *and* to exaggerate Spinoza's distance from the theological tradition generally.

Rather, in part V of the *Ethics* we learn that the divine nature, which of necessity does all that it can do, contemplates its own perfection eternally.[402] This self-recognition of the divine perfection is the divine self-love.[403] As God cannot pass to a greater or lesser perfection, God's self-contemplation does not yield the *affect* of "joy," and consequently the *affect* of "love" is not attributable to God either.[404] Instead, the self-contemplation that God enjoys is an infinite intellectual love, and cannot be anything less than the eternal blessedness which accompanies the self-apprehension of perfection-itself.[405] Our participation in

398 *GL* §168.1; *The Christian Faith*, 732–33.
399 Lamm, *The Living God*, 214.
400 Lamm, *The Living God*, 215.
401 Spinoza, *Ethics*, VP17, P36C.
402 Spinoza, *Ethics*, VP32C.
403 Spinoza, *Ethics*, VP35. See Augustine, *De Trinitate*, ix, 1.
404 Spinoza, *Ethics*, VP17.
405 Spinoza, *Ethics*, VP33S.

this love is also our blessedness. And as God is the *cause* of our blessedness, that is, of our participation in the eternal self-apprehension, "From this it follows that insofar as God loves himself, he loves men, and consequently that God's love of men and the mind's intellectual love of God are one and the same."[406] The eternal self-love of God is God's glory, and our participation in that love our salvation.[407] God is and does all this of absolute necessity. Clearly, Schleiermacher's alteration of this doctrine is not a dramatic departure. In adopting Spinoza's account, he merely adds the notion that the order of the world is perfectly conducive to this end, thus rendering it metaphysically necessary *and* teleologically ordered.

In this context, Schleiermacher's treatment of the divine wisdom receives its full import. Here he bears quoting at length:

> Even in human affairs the primary work of wisdom is correctly and completely to outline the idea which the work of art is to embody, so that actions proper are only traceable to wisdom in proportion as both in the context of a man's life and in themselves they can be regarded as works of art or parts thereof; and he would be the most perfect man of all whose plans for works or actions formed a complete whole of self-communicative presentation. Similarly the divine wisdom is nothing but the Supreme Being viewed as engaged in this absolute (not compositely, but simply and originally, perfect) self-presentation and impartation.[408]

Not only ends, but means, are the measure of a person's life. In the most perfect human life, ends and means would form a coherent whole that, as a whole, express that life completely. So too with the divine life, with the addition that such action not only forms a coherent whole, but that the result is not a composite activity, but a simple and complete self-expression.

And so, Schleiermacher warns that, no *contrast* of ends and means is admissible. Again, he says that:

> Every human artwork is the more perfect, the more it rises to the notion that ends and means do not fall into this contrast [*Gegensaz*] within it, but all just relate as part to whole, and means only lie outside it. In a still higher degree of perfection, this manifestly applies to a complete human life. How then could it not be that the divine wisdom did not exclude the contrast of end and means even more completely?[409]

---

406 Spinoza, *Ethics*, VP36C.
407 Spinoza, *Ethics*, VP36S.
408 *GL* §168.1; *The Christian Faith*, 733.
409 *GL* §168.1; *The Christian Faith*, 733. Translation revised.

Like a perfect painting, the perfect life can be no different than it actually is, either as a complete whole or in its arrangement. Every subject is formed by myriad brush strokes. And every brush stroke is beautiful because of its relation to the rest and to the whole they all comprise. Each exists for the other, and all for the whole. The painting cannot be what it is supposed to be without the color and line that make it up; neither can the paint be what it is supposed to be without its careful arrangement.

Yet, the painter is not the creator of his paints and brushes, but their employer. In contrast, the denial of composition in the divine self-expression implies the denial of the use of external means. And so, Schleiermacher continues:

> There is nothing outside the world which could be used as means; all things within it, rather, are so ordered that viewed in connection with one another they stand related as parts to the whole; while every particular in itself is so entirely both things—means and end—that each of these categories is constantly abrogating itself and passing over into the other.[410]

Nothing in the world is mere means, but always both means and end. As Leibniz explains: "For the wisest mind so acts, as far as it is possible, that the *means* are also in a sense *ends*, that is, they are desirable not only on account of what they do, but on account of what they are."[411] Further, should something supposedly outside the world be used as means, it would enter the sphere of mutual interaction that *is* the world. And so, it would not be outside the world after all. Nothing but the world can be the means of the world's becoming what it is for; the world is so ordered that, like the perfect painting, the arrangement of its parts is as perfectly determined as the whole those parts comprise.

The world as divine artwork is a combination of teleological ordering and the mutual determination of all things. Because, again through conceptual containment, any one thing can only be what it is given, the totality of its relations to everything else, everything in the world is so exhaustively and mutually determined that, down to the very last detail, the totality of finite things, the whole of which they are a part, and the ultimate end for which that whole exists, cannot be otherwise without contradiction. And, because Schleiermacher denies divine transworld identity (because he denies that there are any other possible worlds for God to be God *in*), God is not exempt from this exhaustive system of relations—though the relation of things to God is always one of absolute dependence. Schleiermacher's reconciliation of teleology and necessity turns out to be *automatic* when, following Spinoza, one denies genuine possible worlds, but

410 *GL* §168.1; *The Christian Faith*, 733.
411 Leibniz, *Theodicy* §208; 257.

following Leibniz and the theological tradition, holds to the world's teleological character as ordered not to a divine end of *need*, but to the natural self-communication of divine goodness.

The World is the Absolute Revelation of the Supreme Being[412]

The coherence of teleological and absolutely necessary divine action permits Schleiermacher to follow Spinoza in claiming that God is better known through investigation of the natural order than through the miraculous. Moreover, the compatibility of necessary and teleological action allows him to *combine* Leibniz's argument against miracles with Spinoza's. If wisdom and necessity are compatible, so are arguments which rely on them as premises. And, in fact, if God acts both wisely and of natural necessity, and those two turn out to be one and the same, then Schleiermacher's arguments against miracles from wisdom and power are actually two aspects of a single account, and so neither can be dismissed without accounting for the other. The consequence of both arguments is that absolute miracles simply don't happen. And obviously the non-existent is not worth investigating. Of course, both do allow for *presently* unexplained phenomena. But these cannot be absolute miracles since, in principle, they are caused, and thus can be explained, through the natural order. They are simply events of whose causes we are presently ignorant. Since a thing is called miraculous only through ignorance of its causes, knowledge of God—the cause of the world—is better gained from other, better understood, occurrences.

Arguments against miracles are, however, only stepping stones in Schleiermacher's larger argument. If miracles are impossible, their impossibility simply cements the status of the natural order as that through which God exclusively acts. But, following Spinoza, Schleiermacher argues that not only knowledge of God's *existence* but knowledge of God's *essence* is gained through knowledge of the natural order. This stronger claim requires not only that whatever God does is done through the natural order alone, but also that what God does through the natural order is absolutely necessary. That final step in the argument is what Schleiermacher wins through Spinoza's denial of possible worlds. And,

412 "*Die göttliche Weisheit is der Grund, vermöge dessen die Welt als Schauplaz der Erlösung auch die schlechthinige Offenbarung des höchsten Wesens ist, mithin gut.*" *GL* § 169; *The Christian Faith*, 735.

that is precisely what Schleiermacher argues for when he joins Spinoza in denying the world's merely hypothetical necessity in favor of its absolute necessity.

The transition from hypothetical to absolute necessity only requires the minimal addition that the world's existence, and God's creation of it, are predicates without which the concepts of neither God nor the world could be understood. The minute nature of this change is important. As noted in Chapter 3, Leibniz is already committed to this world being necessarily as it is on the condition of its being willed. That is, Leibniz does not question whether *this world* could be other than it is. Should this world be different, it would not be this world. Every detail of this world is necessarily included in the concept of *this* world. At most what Leibniz allows—at least in his correspondence with Clarke—is that a world other than this is hypothetically possible. But, whether on the condition of this world being willed or not, the concept of this world nevertheless already includes the totality of its predicates, apart from its existence.

Leibniz's position ties the internal necessity of this world to all that happens in it. And, for both Leibniz and Schleiermacher, that includes redemption. In consequence, nothing about the world could be other than it is without changing (and so negating) Christ, his work, and his church. *This* world *necessarily* includes redemption through Christ. As any other world would not be a version of this world, but a different world entirely. So too, any other world would not include a version of Christ, but someone else altogether. Redemption in Christ, seen through the perfection of divine wisdom as contained in the concept of this world, is absolutely necessary to the concept of this world along with all its other predicates. Only by an abstraction can this world be understood apart from its redemptive purpose.

Although the world is teleologically ordered for redemption in Christ, the necessity by which the world is ordered to redemption is not a different necessity than that by which the rest of the world is ordered. That is, while one cannot investigate this world, on pain of abstraction, apart from the redemption accomplished by Christ, so too one cannot investigate that redemption apart from the rest of the world. This is crucial for consideration of the place of natural science for Schleiermacher. *Everything* in the world is included in its concept. The complete comprehension of the world requires the understanding of everything within it, not just Christ, though not less than Christ either. And since the concept of Christ himself includes all *his* predicates, and those predicates include his infinite relations to the natural order, one cannot even completely comprehend Christ without comprehending the world of which he is a part. Although Christ's redeeming work is the end for which the world is ordered, this end cannot be abstracted from the world as means without violating the perfect unity of the world as divine artwork.[413] And while no such comprehensive knowledge of the world is *practically* possible, ideally there is no natural science without Christology, and no Christology without natural science. "Christ therefore was determined as He was," Schleiermacher says, "only because, and in so far as, everything as a whole was determined in a certain way; and conversely, everything as a whole was only so determined, because, and in so far as, Christ was determined in a certain way."[414] The internal necessity of this world follows from the world's concept alone, and that includes its ordering to redemption.

However, as we have seen, Schleiermacher further argues that not only are all the predicates of the world included in its concept, but that this and only this world's existence, and God as creator of it, are also included in the concept of *God*. This one additional step—and only this one more—renders *this* world and all its contents essential to God. In so arguing, Schleiermacher completes the logical circuit that Spinoza sought in his argument against divine transworld identity. Just as for Spinoza, the consequences of Schleiermacher's claim for the investigation of the natural order are tremendous. For not only is the world the exclusive means through which God works, but also the *essential* means. God not only necessarily wills the divine essence as end, but necessarily wills that end *and its means* of natural necessity. The relation between divine essence, divine end, and divine means is thus rendered deductive.

This crucial final step allows the application of the transitive property (if $a = b$, and $b = c$, then $a = c$) to the divine essence, end, and means. When made explicit the consequences of this claim are extraordinary. If the divine means are absolutely necessary to the divine end, and the divine end to the divine essence, then the divine means are absolutely necessary to the divine essence. Since whatever means God uses are necessary to the divine essence, and since God uses no means other than the natural order, the natural order is necessary to the divine essence. As the investigation of the natural order is an investigation of the divine means, and those means the absolutely necessary expression of the divine essence, *the investigation of the natural world, (i. e., natural science), is, by the transitive property, the investigation of the divine essence.*[415]

---

413 See *GL* §94.3; *The Christian Faith*, 388–89.
414 *GL* §120.3; *The Christian Faith*, 555.
415 In this case, the investigation of the natural world is the investigation of the divine essence as the investigation of the divine essence's effects. The inference from divine effects to God as cause is, of course, perfectly traditional. But by insisting on absolute, not merely hypothetical, necessity, Schleiermacher makes this an inference from necessary, but dependent, effects to necessary, but independent, cause, such that if anything in the world were to be other than it is it would imply a contradiction of the *divine essence.*

That the investigation of the natural world is the investigation of the divine essence makes plain why Schleiermacher claims that the *world* is the absolute revelation of the Supreme Being.[416] It is not merely Christ who expresses the nature of God—though Christ is indispensable for rendering that expression determinate. Rather, it is the whole of the natural order, the *entire* concept, including Christ, which is the self-expression of God. As the perfect work of art, it is the perfect life expressing its essence in a unity of means and ends, so the world as a whole—and only as a whole—is a supremely beautiful divine artwork and so, as the perfect self-presentation and impartation of God, is also the absolute revelation of the divine essence.

Purpose, Necessity, and Evil

As grand as the necessity of the world as the divine self-expression is, that doctrine might appear to wither in the face of objections from evil. If the world *in its every detail* is necessarily as it is, then so is every evil. And since what happens in the world is an expression of the divine essence, evil appears necessary to the divine essence. Absent some defense of the necessity of evil, Schleiermacher's account would be set back by its own advance. However, he responds to objections to his account on several points. The first of these is that he makes God the cause of evil; the second is that he makes God the necessary cause of evil and so evil itself necessary.

In response to the first charge, that he makes God the cause of evil, Schleiermacher not only concedes the point, but insists on it. God is the cause ("author" [*Urheber*]) of all things, and has "ordained" all things—including evil.[417] Against his objector, he presses back that should God not be the cause of evil, evil would need to be explained otherwise.[418] And since, along with both Leibniz and Spinoza, he holds that every explanation must terminate in a self-explaining explanation, should evil not be explained by a more ultimate explanation, namely God, then it would terminate in some other self-explaining explanation; and so, evil would have its reason for existence apart from God. Should God not be the cause of evil, evil would be self-caused, and cosmic dualism would result. One of

416 "[…] *die Welt als Schauplaz der Erlösung auch die schlechthinige Offenbarung des höchsten Wesens ist* […]" *GL* §169; *The Christian Faith*, 735.

417 *GL* §§48, 79, 82; *The Christian Faith*, 184–89, 325–26, 338–41.

418 Even an account of evil by deficient causes would eventually need to terminate in God as the cause of the lack which is the condition of the possibility of evil. See *GL* §81; *The Christian Faith*, 330–338.

the very few heresies Schleiermacher names is the "Manichaean" type, which admits of the self-subsistence of evil,[419] and any account that does not hold that God is the cause of evil would conflict with that most basic commitment to monotheism.

This defense-though-offense strategy does not, however, address the specific problem of the *necessity* of evil. To that problem Schleiermacher has a more subtle response. First in his defense, it should be noted *how* evil is necessary: evil is, in one sense, only conditionally necessary. That is, evil exists only because of something else, something extraordinarily good, that, without evil, could not be. Evil cannot be isolated from the world at large. "The fact is rather," says Schleiermacher, "that the very same activity or condition of a thing by which it enters on the one hand into human life as an evil, on the other hand is a cause of good, so that if we could remove the source of life's difficulties the conditions of life's progress too would disappear."[420] That is not necessarily to say that every evil *is* a good, and so to deny evil *tout court*. Rather, it is to say that the conditions of every evil are also, however directly or indirectly, the indispensable conditions of some good that could not *be* without evil.

However, it should also be immediately reiterated that *this kind* of conditional necessity is not any less *logically* necessary than absolute necessity. Both are still absolutely necessary in that their opposite implies a contradiction. Instead, as above, Schleiermacher, following Spinoza, distinguishes the *kind* of necessity ("any change violates the concept," versus, "its opposite does not imply a contradiction") from the *reason* for a thing's being necessary.[421] This is most obvious in his doctrine of the world's dependence on God, which includes both the logical necessity of the world in God, but, at the same time, the absolute dependence of the world on God. Likewise, for evil, it is logically necessary, as all things are, in that it cannot be other than it is without violating the concepts of world and God. However, its *reason* for existence is not itself. Its reason, moreover, is not merely God as efficient cause but also God as final cause. Evil is willed as a condition of some great good: namely, the self-communication of the divine essence through Christ and his church. Without evil this great end could not be accomplished perfectly. And since any end but this, and any means but perfect means, violate the concept of God, any world where evil does not occur exactly as it does occur implies a contradiction. "Nothing remains," Schleiermacher concludes, "but on the one hand to attribute the divine cooperation equally to every-

---

419 *GL* §§22.2, 80.4; *The Christian Faith*, 98, 329–30.

420 *GL* §48.2; *The Christian Faith*, 187.

421 This helpful distinction comes from Griffin, *Leibniz, God and Necessity*, 73.

thing that happens, and on the other to maintain that evil *as such* is not ordained by God, but only as related to the good and as one condition of it."[422]

To say that the world is perfectly ordered is *not* an account of this world as the best of all possible worlds. Not only has Schleiermacher dismissed the idea already, but he has done so with arguments that literally make any world but this one inconceivable. If the *concept* of God includes the divine ends and means, and those ends and means include this world and everything that will happen in it, then it is absurd to entertain a different world. It is a violation of a geometrical truth that the world could be otherwise. To entertain a different world than this is like trying to think of a triangle with four sides. Appeals to the imagination are of no avail. The ability to imagine this world differently, or rather, another world than this, necessarily falls short of the whole to which it is accountable. Any imagination will be an abstraction, an incomplete thought. It will necessarily divorce things from the causal network in which they exist as they do. And since everything is what it is only in virtue of everything else, to abstract even one thing from the world as a whole is to change it, and the world along with it. It is to unthink the world.

Like a geometric object, the world necessarily is such as it is, both as a whole and in its parts. And yet, as with geometrical analysis, the formal necessity of a thing might be known without a complete analysis of the concept *exhibiting* this necessity. For instance, even very young children can be taught what a triangle is. It is one kind of thing and not another kind because that is what the word means. Other shapes are known by other words. Children can see that triangles, and only triangles, are closed shapes with three and only three sides and that a purported triangle with four sides is, in fact, not a triangle at all. It is, however, more advanced, a task for older students, to exhibit the necessity of the analytic truths of this concept through the measurement of angles, the ratios of sides, etc. These are, of course, not really two separate tasks, but the latter a more complete version of the former. Once these analyses are performed, the same concept is still as necessary as before, and all its nascent meanings still contained within the concept. What has been added is not new information but a new understanding, a demonstration, of the internal necessity which was already implied in the concept.

Schleiermacher, following Spinoza, has argued that the world necessarily is such as it is. And while he has shown *that* the world must be as it is necessarily, he has not yet *exhibited* its internal necessity. This task is crucial, and yet it is not the theologian's alone. Instead, it is also the task of *natural science* to exhibit the

---

422 *GL* §48.3; *The Christian Faith*, 189. Emphasis added.

internal necessity of the natural order ever more completely. Since nothing in this world is completely understood unless it is understood in its manifold necessary connections to everything else, the interests of natural science and theology meet at the same point. And, importantly, since evil is not outside the natural order, but included in it, the task of exhibiting the internal necessity of the natural order is one and the same with the task of exhibiting the necessity of evil. Thus the *demonstration* of the internal necessity of evil is not the theologian's task alone, but also the natural scientist's. That is, the exhibition of the internal necessity of the world is achieved by specifying its contents and their relations so completely that to change one thing about the world would change everything else. But since evil is a feature of the natural world, specifying the world's necessity and evil's necessity are one and the same activity. Since natural science is tasked with the former, it is tasked with the later. Of course, both are infinite tasks, and so in practice unachievable in this life. Nevertheless the conditions of their completion are identical.

Moreover, because Schleiermacher holds that the world is teleologically ordered, the giving of good reasons why evil should be is not a separate task. Instead, it too follows analytically from the ordering of goods and the exhibition of their necessity. That is, any teleological order will, by definition, be ordered to some ultimate good. And, the means to that good and their conditions will be *for* the accomplishing of that good. If *actually so*—that is, if the proposed ultimate good turns out to be genuinely good and genuinely ultimate—then an order which genuinely contributes to the accomplishment of that ultimate good, and which cannot be accomplished better, is, by definition, also good. Or, put differently, to rationally act by some means for some end *minimally* implies that the end for which the means are means is worth the means. Otherwise, one would intend a greater good for the sake of a lesser good, which is absurd. Since Schleiermacher holds that the order of the world is absolutely necessary, it cannot be better than it is. And since Schleiermacher's argument from wisdom is concerned with showing that the natural order is not only a necessary condition of the accomplishment of the divine end, but also a completely sufficient one, the world is *precisely* that which alone contributes to the accomplishment of the ultimate good for which the world was made. In both senses, that the world is sufficient for its end, and that it cannot be better than it is, the world is perfect. Any remaining questions, therefore, can only be addressed to the worthiness of the ultimate good.

Since that ultimate good is union with God, not only can no better good be assigned, but that good is, in fact, infinite. There is no proportion between it and other, lesser goods. That then only leaves the question of whether or not all *individuals* were made for that end, and, if so, if they will, in fact, come to enjoy the end for which they were made. The sufficiency, and so perfection, and so goodness, of the world as a whole might well be impugned by the failure of some individuals to enjoy this ultimate good. It is in regard to this final question that I have argued elsewhere that Schleiermacher not only does, but *must*, on pain of incoherence, hold to universal salvation; that is, to the actual accomplishment of the impartation of the divine essence to every individual, and so to the supreme good of the eternal enjoyment of fellowship with God which follows from it.[423]

The most powerful argument that Schleiermacher not only did, but *must* hold to some form of eternal life, and that it *must* ultimately be enjoyed by each and every individual is the teleological perfection of the world taken to its natural conclusion. On Schleiermacher's account, the world is created and ordered precisely for its final end, which is our redemption. Since the world was created *just for this end*; and since, by the divine wisdom, it is perfectly ordered to its end; and since, by the divine power, its end shall surely be attained, the ultimate end for the world cannot fail to be fulfilled.

Moreover, Schleiermacher affirms, in no uncertain terms, that this end is *one and one only:* the redemption of all through Christ. He flatly denies double predestination.[424] He flatly denies reprobation as anything other than a *temporary* passing over.[425] He even doubts that annihilation is sustainable in light of the single fore-ordination to blessedness in Christ.[426] The perfectly ordered world perfectly attains its single end.

However, another very different objection might suggest itself. Perhaps the divine end is one, but, because Schleiermacher has ruled out miracles, it is not possible for *anyone* to enjoy eternal fellowship with God. This second argument, if apparently more plausible, is weaker still. For Schleiermacher's argument against absolute miracles from the teleological perfection of the world only works on the supposition that the world is, in fact, perfectly ordered to its end. But this end *is* the redemption of all in Christ. Therefore, it is supremely ironic to argue on the basis of Schleiermacher's denial of the miraculous that eternal life is impossible; it is to conclude that that the world is so perfectly ordered to its end that its end cannot be attained, which is a contradiction. No, Schleiermacher's account of teleological perfection does not exclude the possibility of eternal life, but necessitates it.

423 Daniel Pedersen, "Eternal Life in Schleiermacher's *The Christian Faith*," *International Journal of Systematic Theology*, 13 (2011), 340–357.
424 *GL* §118.2; *The Christian Faith*, 544.
425 *GL* §119.2; *The Christian Faith*, 548–549.
426 *GL* §118.3; *The Christian Faith*, 545.

Since Schleiermacher holds that the natural order is the all-sufficient (i. e. perfect) means that cannot, apart from contradiction, be other than it is, and that certainly and inevitably accomplishes the supremely good end of union with God for each and every individual, the world as a whole, and evil as a feature of that world (since the world cannot *be* without evil), is also worth it. Provided the supremely good end of eternal enjoyment of God is actually attained by every individual, the ordering of the world to that end does not undermine God's goodness.[427]

Lest it be objected that this argument is crassly consequentialist, that evil is somehow justified merely because it is a means to a good end, several points require repeating. The first point simply regards the relation of means to end. As above, any reasonable teleological action will *minimally* always use some lesser means for some greater good. That is, again as above, if an action were ordered greater means *to* lesser good it would not be an example of practical reasoning, but some species of error or insanity. As Candace Vogler has shown, all practical reasoning is "calculative" insofar as it weighs means to ends or parts of activities to the whole activity.[428] And so, if all means-to-ends actions are consequentialist in this calculative sense, then that charge turns out to be trivial since there is, by definition, no teleological account, and so arguably no description of rational action at all, that does not include consequentialist reasoning in this basic sense. Something more is required to reduce Schleiermacher's account to the status of *mere* consequentialism than to point out that on his account God is, in some sense, a practical reasoner. Moreover, as Schleiermacher's elaborate analogy of divine art makes clear, God does not *crassly* use means for ends at all since the world is not only useful, but beautiful, not only a fitting means to a good end, but also a work befitting the artist whose character finds expression in it.

The second point which deserves repeating is that, again, "evil *as such* is not ordained by God, but only as related to the good and as one condition of it."[429] That is, even though evil does not fall outside the divine causality, it is never a discrete object of divine causality. God, in fact, ordains *nothing* discretely, and so,

---

427 Leibniz seems to have noticed the usefulness of universal salvation for answering difficulties posed by evil when he claimed that, "An Origenist who maintains that all rational creatures become happy in the end will be still easier to satisfy. He will say, in imitation of St. Paul's saying about the sufferings of this life, that those which are finite are not worthy to be compared with eternal bliss." Leibniz, *Theodicy* §211; 260.

428 Candace Vogler, *Reasonably Vicious* (Cambridge, MA: Harvard University Press, 2002), 22–23.

429 *GL* §48.3; *The Christian Faith*, 189. Emphasis added.

certainly not evil. Instead, God only ordains the whole as a whole, along with all the particulars that world entails. This point is crucial. For Schleiermacher's claim is not that God ordains some evil *simpliciter* as the condition of some good. No, God only wills any one thing on the condition of *everything* else. Rather, Schleiermacher's claim is that God wills the world as a condition of some good, and evil only as a necessary condition of that world, where evil as a necessary condition is understood not only as a prerequisite but also as an implicate. This doctrine has far more to do with Schleiermacher's subscription to exhaustive conceptual containment than any consequentialism. For, if any complete concept includes *all* its relations, then when God wills the concept of a world at large, God necessarily wills those relations. So if, for example, God wills a world which brings about humans to be redeemed, then God must will the kind of world where humans can come to be, and that means *this* world in all its specificity. As this world necessarily includes evils great and small, from war, to hurricanes, to bee stings, without which the world as it is cannot be conceived, God wills those evils insofar as God wills the world, but *only* as a condition of willing a world which, as a whole, is perfectly sufficient to the divine end of redemption.

However, it might be objected at this point that that conclusion is begging the question, that whether the world could be *better* than it is is precisely what is being pondered. And so, it might be objected, any appeal to the world's necessity in support of Schleiermacher's conclusion is inadmissible. Such an objection, however, would be mistaken. The objection naturally suggests itself because the teleological ordering of goods follows the necessity of the best. However, Schleiermacher does not appeal to *this* necessity when arguing for strict conceptual containment in both the world and God. Instead, Schleiermacher's initial appeal is based on Spinoza's argument for *geometrical* necessity, which is itself not dependent on explicit appeals to goodness and wisdom. Recall instead that Spinoza concludes to a being whose essence involves existence from certain necessary and sufficient conditions of explanations, namely: that everything which is not explained through itself is explained through another; that infinite regresses are inadmissible; and so the explanatory series necessary terminates in a self-explaining explanation. And it is from an analysis of a self-explaining explanation, a being whose essence involves existence, that he concludes that determinate effects follow necessarily from determinate causes, and so on. It is this second *reason* of necessity that generates the impossibility of the world being other than it is. And since this necessity also contradicts divine transworld identity, the traditional theodicy question changes correspondingly. It is no longer a question of whether God exists, given evil. Instead, the question is: given that this world is as it is and can be in no other way, and that there is, in fact, an explanation for the world which terminates the regress, *how is that self-explaining explanation (and its actions) best described?* Though ultimately identical, the necessity of the world precedes its goodness in the order of presentation, and so no question is begged when necessity is appealed to in answering objections to the world's goodness. Instead, the combination of both the Leibnizian necessity of the best and Spinoza's geometrical necessity become, in Schleiermacher's hands, two mutually supporting parallel arguments.

The Christian Faith concludes that the *Whence* of the world into which we are inquiring gives itself through Christ's God-consciousness, and so is known as love; since the world is ordered, it is ordered to this love by wisdom; and, finally, that as the work of almighty-wise-love, the world is therefore *good.*[430] *The Christian Faith* itself follows its own argument in structure. It begins with questions of the necessary and sufficient conditions of the *Whence* of the world (i. e. of an explanation which explains itself, that is, a necessary being, a being whose essence involves existence), and proceeds to more fully describe that being in valued terms. Note though that this *does not* entail a fact-value divide. The only reason the goodness of God is not obvious from the beginning is that God's action is incompletely specified. It is precisely the determination of divine action which the redemption accomplished in Christ provides. Put another way, it is not the case that a Schleiermacher admits of a genuinely ontologically indeterminate *Whence*, only that the determinate nature of the one *Whence* is (temporarily) incompletely described. But, in this case, it is important for Schleiermacher to proceed in just this way precisely because, among other reasons, it avoids begging the question in regards to the world's necessity.[431] The conclusion, of course, turns out to be one and the same thing: God's and the world's goodness follow, of necessity, from their concepts adequately comprehended. But, by structuring his presentation as he does, Schleiermacher settles the question of necessity before specifying the nature of that necessity in valued terms. The order of presentation in *The Christian Faith* does such important work for Schleiermacher that one can see why he was reticent to reorder the work in spite of the widespread misunderstanding it produced, and still produces.

Schleiermacher's defense of the Christian faith in light of evil does not purport to demonstrate the coherence of God and evil in some generic religion and

---

430 *GL* §169; *The Christian Faith*, 735.

431 It should be noted that the attributes of omniscience and omnipotence appear before *any* Christological content. And while Schleiermacher clearly states that a generic omnipotence has nothing to do with the God of Christianity, the later Christological modifications to the omni attributes do not negate them. Rather, Christology renders principles determinate by specifying their content.

indeterminate world. Instead, it assumes the specification of divine action as known through the redemption accomplished by Christ. Granted the *telos* of divine self-communication (i.e. love) which that entails, the necessarily ordered world is ordered wisely, and so is good.

Chapter 7 Conclusion: The Essential Identity of Ethics and Natural Philosophy

> "But although the annexation of that philosophy which concerns the theory of Nature to this work appears to overreach his original plans, still not only is the necessity for it declared in his own words, but even the first outlines drawn according to which they are to set to work upon this subject. [...] And thus the manner in which the *Timaeus* connects itself with the books of the *Republic* is a declaration of the essential identity of ethics and natural philosophy."
>
> - Schleiermacher, *Introductions to the Dialogues of Plato*[432]

We began this investigation with the question of what Schleiermacher's proposal for an eternal covenant between faith and science amounted to. In this last chapter I revisit this initial question, elaborate upon what I take to be its implications, and entertain objections to my position. Some of my conclusions differ from venerable and popular interpretations of Schleiermacher's thought. Throughout this chapter I rely on the arguments of previous chapters as warrant, and so I ask the reader to return to those chapters if in doubt of my claims or their consequences.

In the first chapter, I argued that Schleiermacher's proposal could not have been a purely political arrangement or methodological starting point. Instead, I claimed that the eternal covenant was the result of argumentation. The eternal covenant is not a step in these arguments, but a consequence of them. By now that should be clear. Instead, Schleiermacher's arguments proceed *ex hypothesi* from commitments he and his fellow theologians already share: from the distinction between independent and dependent being, from divine power, and from divine wisdom. In each argument, Schleiermacher began with premises that are both traditional and highly plausible assumptions on the part of theologians, and yet which led to a controversial account of divine action: namely, that God only ever acts through secondary causes, and never supplements, emends, intervenes upon, or specially co-operates with created things beyond preserving them as beings with sufficient innate powers to act.

---

432 Schleiermacher, *Introductions to the Dialogues of Plato,* trans. William Dobson (London: 1836), 413.

DOI 10.1515/9783110542301-007

That is the first step in securing the basis of the covenant. By eliminating all notions of divine action apart from one, Schleiermacher guarantees that everything that happens in the world is a natural event with natural causes, amenable to scientific investigation. But that consequence is not yet a covenant between faith and science, not yet a friendship in pursuit of a common purpose or shared goods. Had Schleiermacher stopped here, theology would have gained only the very modest good of abandoning unworkable notions. And, for its part, natural science would also have enjoyed only the very modest good of working without theological impediments. These are gains, to be sure, but only with the greater benefits to each can the covenant between faith and science can be rightly so-called.

These larger gains only come with Schleiermacher's second step. The second step in securing the basis of the covenant is to connect the world to the divine essence through the chain of determinate causes. By denying the genuine possibility of any world other than this one, Schleiermacher renders this world in all its detail not only hypothetically necessary, but absolutely necessary, and thus intrinsic to the very notion of God.[433] In so doing, Schleiermacher secures the *world*, necessarily including (but not restricted to) Christ, as the absolute revelation of the divine essence.[434] It is the world as the absolute revelation of the divine essence, relying on the first step (that God only ever acts through secondary causes[435]), that makes the eternal covenant both a covenant and, in a meaningful sense, eternal.

The world as the absolute revelation of God, as we saw in Chapter 6, makes the investigation of the natural world the investigation of the divine essence by virtue of the transitive property.[436] In consequence, whenever we learn of something being natural, being a part of our world, we finally and ultimately learn something essential about God. And the benefits of this approach are not only benefits to natural science or only benefits to theology, but benefits to both. For both theology and science, the main benefit of Schleiermacher's various arguments for securing the absolute necessity of the world as it is, including Christ, is to eliminate any possible rift between God's being and God's activity, any possible gap between the appearance of God and God's reality. On the

433 See Chapter 5.
434 See Chapter 6.
435 See Chapters 3 and 4.
436 The transitive property is that by which if a=b and b=c, then a=c. It is a fundamental principle of reasoning, like non-contradiction, which is required for many ordinary inferences. In Schleiermacher's particular use this warrants the inference from the world to God as necessary, but dependent, effect to necessary, but independent, cause. See Chapter 6.

part of natural science, this implies that, when scientists investigate the natural world, they are inquiring about ultimate reality—only indirectly, as the world considered as the means of divine wisdom, of physics.[437] On the part of theologians, this implies that, when theologians make claims about human nature, human action, and human and divine purposes in history, they are making claims about nature, since there is no finite thing that is not part of the world, and everything that occurs in the world can be explained by nature as a whole—only indirectly, as the teleological ordering of the world, of ethics.[438]

---

437 By "physics" I understand Schleiermacher's treatment of the notion in his *Introductions to the Dialogues of Plato* where the term stands for nature most broadly, not in the narrower sense as it is sometimes used in the Introduction to Schleiermacher's *Lectures on Philosophical Ethics* where the subfield of physics is defined as "the recognition of the essence of nature" and is distinguished from "knowledge of nature" as "The contemplative expression of finite being, inasmuch as it is nature." Nevertheless, Schleiermacher occasionally also uses "physics" in the *Lectures* in this same broader sense. For example, he explains, "The highest unity of knowledge, expressing both domains of being in their interrelatedness, as the perfect [mutual] permeation of the ethical and the physical and the perfect simultaneity of the contemplative and the experiential, is the idea of 'world-wisdom.'" That is, "both domains" in this case are the two domains of highest opposition—"the ethical" and "the physical"—which, in their perfection, constitute the highest unity of knowledge, not the specific subfields that go by the same name. Moreover, Schleiermacher also uses "ethics" and "physics" in the broadest sense in the gloss to this same proposition where he explains, "This [*Weltweisheit*] is the perfect reproduction of the totality of being, just as that [the totality of being] is the direct image of the highest being. It can never be complete, however, as long as ethics and physics exist as separate sciences." See Schleiermacher, *Introductions to the Dialogues of Plato,* trans. William Dobson (London: 1836), 299; Schleiermacher, *Lectures on Philosophical Ethics*, ed. Robert B. Louden, trans. Louise Adey Huish (Cambridge: Cambridge University Press, 2002), Introduction, final version, §59; 150–51.

438 By "ethics" I also understand first and foremost Schleiermacher's treatment of the notion in his *Introductions to the Dialogues of Plato* as regarding the good and of explanations in terms of the good. Schleiermacher, *Introductions to the Dialogues of Plato,* 298–99.

This meaning of "ethics" is, however, perfectly compatible with Schleiermacher's meaning in the various versions of his Introduction to the *Lectures on Philosophical Ethics* where "ethics" is not understood as the specific subfield of *Sittenlehre*, that is, the specific "science of morals," but rather as the second of the highest main sciences—ethics most broadly, that is, "Reason." In the 1812/13 version, Schleiermacher provisionally defines this second main science (what the *Introduction to the Dialogues of Plato* calls "ethics") as "the life of reason" which "must therefore encompass and catalogue all human action." At the highest level, "Ethics is thus the depiction of finite being under the power of reason, i.e. viewed from that aspect where, in the interrelatedness of oppositions, reason is the operative principle and reality that which is acted upon." Schleiermacher, *Lectures on Philosophical Ethics*, Introduction and doctrine of goods, 1812/1813 version, §§12, 16, 28 Lemma 10; 4–5.

The connection between Schleiermacher's use of the term in the *Lectures* to his use in the commentary on Plato's dialogues becomes clearer in the cover notes added to the final edition of

What the eternal covenant does is to reunite their abstracted aspects of investigation—divine ends and means, love and wisdom, ethics and physics—into a single account, a unity that befits the one *Whence* of whom the world is a perfect and undivided appearance.[439]

The cooperative union of theology and natural science answers why the covenant can be a common pursuit of shared goods. Theology and natural science *need* each other to offer the complete account that the unity of the world demands. One without the other is a vicious abstraction which cannot, in principle, hope to attain to knowledge of the natures of things, either natural or divine.[440] And this is because their *objects* are, in fact, the same—both the world, and the divine nature—only indirectly, each in terms of the other. Theology and natural science are engaged in a cooperative pursuit of shared goods *because the goods they pursue are, in the final analysis, one and the same.*[441]

If the investigation of the natural world, as an undivided whole, under the twin aspects of divine ends and means, and as a unity of both ethics and physics, makes clear why the eternal covenant is properly called a covenant, then the absolute necessity of the world in its every detail, and the absolute necessity of

---

his Introduction where Schleiermacher explains that, "Everything existing for its own sake—which is, at one and the same time with reference to its whole sphere, something which generates and is generated—is something organized, which has become ethical; it is a good. Thus the totality of what exists in ethical terms for its own sake is the system of goods and the organism of reason, so that ethics is the doctrine of the highest good." Schleiermacher, *Lectures on Philosophical Ethics*, Introduction, final version, n. 26; 167. The notion of "reason" Schleiermacher has in mind regards the good and ethical explanations in terms of goods. It is the reason proper to intelligence which Anaxagoras failed to muster.

439 "Thus all knowledge can only be both complete and perfect *if it can be taken as a whole.*"; "The highest image of the highest being, however, and thus also the most perfect conception of the totality of all definite being, is the complete permeation of nature and reason." And again, "This [*Weltweisheit*] is the perfect reproduction of the totality of being, just as that [the totality of being] is the direct image of the highest being. *It can never be complete, however, as long as ethics and physics exist as separate sciences.*" Schleiermacher, *Lectures on Philosophical Ethics*, Introduction, final version, §§6, 48, 61; 137, 147, 151. Emphasis added.

440 "The highest knowledge, however, is also only perfectly understood if what is especial and subordinate to it is perfectly understood." Schleiermacher, *Lectures on Philosophical Ethics*, Introduction, final version, §5; 137.

441 The perfection of the sciences in identity as the highest knowledge is equivalent to complete knowledge of God: "The highest knowledge is, however, not at all the denoting of a definite scope either, but the indivisible and unmultipliable expression of the complete and absolute highest being which is its equal, just as the highest being is the indivisible and unmultipliable expression of the complete and absolute highest knowledge which is its equal." Schleiermacher, *Lectures on Philosophical Ethics*, Introduction, final version, §30; 143.

the world to the divine essence, make clear why the covenant can rightly be called eternal. When, by the transitive property, the divine essence is known through the world, it is not merely known as transitory. Rather, because there is no discrepancy between the perfect artist and its art, the world is a perfect revelation of the eternal nature that created it; not merely a moving image, but a moving image *of eternity*. And because the basis of the covenant is the unity of the divine artistry, the covenant itself, like the divine nature, can neither come into being nor pass away.

On my reading, Schleiermacher meant the eternal covenant to be persuasive, particularly to theologians, but not only to theologians. In order to argue as he does, he could not assume the conclusions that formed the basis of the covenant from the start, but instead had to argue from *some* shared principles and commitments, like the principle of sufficient reason and the principle of monotheism, which, if not universally held, were at least generally plausible and largely agreed upon. Having explored how his argument proceeded as a whole, we are now in a position to revisit the role these principles and commitments played in his case.

As Schleiermacher insists throughout *The Christian Faith*, our consciousness of God and world go together.[442] And yet, they are not simply two different implications of the one God-consciousness.[443] Instead, the ability to apprehend the world *as a world*, and the capacity for redemption, are distinct but mutually informing natural passive potencies. Both are capacities inherent in human nature which must be activated from without: one by interaction with the world, the other by the power of Christ's spirit. Consciousness of the world grows through the mutual interaction of relatively dependent beings,[444] while consciousness of God grows in proportion to consciousness of the world,[445] and is completed in the specific consciousness of redemption through the implantation of Christ's perfect God-consciousness.[446] Since there is no consciousness of *absolute* dependence without the consciousness of the relative freedom and relative dependence which is part and parcel with awareness of the world, the activation of the natural capacities that are the conditions redemption and natural science are indexed to one another. Increasing recognition of the world as a world goes

---

442 *GL* §34. See also Gerrish "Nature and the Theater of Redemption", 204.
443 *GL* §32.2; *The Christian Faith*, 132–33.
444 *GL* §34.1; *The Christian Faith*, 137–40.
445 *GL* §§46.2, 47.1, 49.2; *The Christian Faith*, 173–75, 178–81, 192–93.
446 *GL* §34.3; *The Christian Faith*, 139–40.

hand in hand with the perfection of the God-consciousness. And so, it is hardly surprising that the two must be *combined* to arrive at the divine wisdom.[447]

Consciousness of the world *as a world* is a precondition of the perfect consciousness of God as almighty-wise-love. Not only is this a necessary *developmental* condition for the consciousness of redemption, but it is also a necessary *explanatory* condition. That is, not only has humanity historically developed an increasingly determinate, and so increasingly complete, sense of the *Whence* of our existence, and not only do individual people develop this consciousness from birth to adulthood, but such a consciousness also serves an explanatory role. Some readers might be suspicious of the claim that the consciousness of God and world is explanatory, but its best evidence is that Schleiermacher's account is a *causal* account. Causes *are* explanations, as are relations of dependence.

And yet, the developing consciousness of both world and God has not only been the development of *an* explanation, but more broadly, of the necessary and sufficient conditions of *all* explanations. It is a formal requirement. Specifically, it is the consciousness of being part of a world that entails that determinate effects be explained by determinate causes; it is the consciousness of being in a system of mutual interaction, and so relatively dependent, that demands absolute dependence as a corollary to explain this system; and it is the consciousness of being absolutely dependent on a single *Whence* that entails the unity of ultimate explanations. Consciousness of both world and God demand both the formal necessary and sufficient conditions under which certain explanations can be expected, and the kind of explanations which would satisfy those expectations. In short, consciousness of world and God serve as explanatory principles.

Principles specify the formal requirements of explanations but don't yet specify their content. *In principle* every effect has a cause, but only determinate content can specify that "this smoke" is an effect, and "that fire" its cause. This is why, on Schleiermacher's account, we cannot have *speculative* knowledge of God: determinate content is needed to specify the exact Christian understanding of the divine nature, and that can only be discovered empirically.[448] However,

447 See Chapters 4 and 6.

448 It should be emphasized that the contrast between speculative and empirical knowledge only regards the order of discovery, *not* the order of being. That is, by the absolute necessity Schleiermacher has insisted on throughout *The Christian Faith*, once we discover said content empirically we know that the divine nature necessarily is and acted one way and no other eternally, that all the content which we had to discover by experience was actually contained in the very nature of God all along. See Chapter 5. It is this same absolute necessity that authorizes

without the principle, the relation between the smoke and fire remain unintelligible.

For Schleiermacher, the consciousness of the world, as a world, functions as a principle in this way. It specifies the kind of explanation which will satisfy, namely, an account of the world as a *system*, without yet specifying its content. When specified, that content will, by definition, turn out to be one thing and not another; moreover, not only the things but the totality of their relations will turn out to be exhaustively determined. That account of the necessary determinacy of comprehensive understandings once again yields an account of the world, as a world and in all its details, which necessarily is such as it is. Thus the branch of human knowledge tasked with the specification of the contents and operations of the natural order, i.e. natural science,[449] is, in consequence, tasked with the exhibition of its internal necessity. And, in fact, the two tasks turn out to be one and the same thing. The specification of the world's contents and the exhibition of its internal necessity is the business of natural science *in principle* because the consciousness of being part of a *world* sets its necessary and sufficient explanatory conditions.

The consciousness of being absolutely dependent on a single *Whence* for our existence is indexed to our consciousness of being part of a world. It too functions as a principle which sets the necessary and sufficient conditions by which the world in turn can be explained. In this case, the necessity of explanations terminates in a *single* self-explaining explanation. Again, Schleiermacher calls this "the principle of monotheism."[450] The self-subsistence of God plays an important role in Schleiermacher's argument where the necessary essence of the divine life begins the determinate causal chain which cannot be other than it is. In this respect, the principle that there is *a single Whence*, a single self-explaining explanation, has already been put to a great deal of work.

These higher principles are significant because, although indexed to the God-consciousness, and so perfected by it, their basis (i.e. consciousness of the nature system) is nevertheless to be found naturally in all people, even if ac-

---

Schleiermacher to claim that the world is the absolute revelation of the divine essence. See Chapter 6.

449 Here I use the term in the common contemporary sense as the study of the natural world, not as Schleiermacher's technical subfield of "physics, or natural science" distinguished from "knowledge of nature" by virtue of being contemplative, rather than experiential. The contemporary sense does not make this distinction, but includes both subfields of physics and knowledge of nature. Schleiermacher, *Lectures on Philosophical Ethics*, Introduction, final version, §59; 150.

450 *GL* §56.1; *The Christian Faith*, 230.

tivated imperfectly and intermittently. And, however they are derived, as principles they logically precede their content. That is, regardless of how one comes to demand, say, a single self-explaining explanation as a terminus to the series, once one does demand such an explanation, every specific explanation must meet the same necessary and sufficient conditions as any other. Irrespective of content, certain formal conditions will need to be satisfied. Further, these explanatory principles admit of more than just true or false, but instead fall on a spectrum of completeness. More stringent explanatory conditions will be more complete than less stringent ones, and explanatory conditions plus determinate content will be more complete still. What all this amounts to is an account of public explanation—specifically scientific explanation—on analogy with pagan virtue. Although indexed to the God-consciousness, and so mutually perfecting, the consciousness of a world and a corresponding demand for certain explanations in principle is at least genuinely (though incompletely) available to anyone. It is these explanatory demands, not restricted to the Christian church alone, which furnish an order to the world as a world, and which *precede* the specification of that order as wise.

Beyond the universally available consciousness of being in a world, the religious self-consciousness furnishes yet more explanatory demands. One far-reaching implication of the "principle of monotheism" is the unity of ultimate explanations. Because a sufficient explanation is a complete one, all explanations must necessarily terminate in a self-explaining explanation. But, that might not obviously imply that all explanations terminate in a *single* common explanation. For example, as we saw above in regard to evil, it is tempting to offer some other independent explanation for evil: a self-subsistent evil from which all other evil is derived. Schleiermacher's commitment, in principle, to a *single* sufficient explanation rules out what in this case amounts to cosmic dualism. But it does not stop there. Rather, Schleiermacher rejects *all* explanations which don't ultimately terminate in a single self-explaining explanation.

The first example of Schleiermacher's insistence on the ultimate unity of explanations, and that which has received the most treatment in this work, is Schleiermacher's rejection of the analytic/synthetic distinction applied to great effect by Kant,[451] and Leibniz's distinction between truths of reason and truths of fact, for that matter. As we saw in Chapter 6, by insisting on the exhaustive determination of all things, even to the point of rejecting divine transworld identity (again, because there are no other worlds for God to be God in), all truth is ultimately analytic because all things ultimately cannot be other than they are

451 See Frank, "Metaphysical Foundations," 22–3.

without contradiction.[452] This does not mean that such conceptual mastery has ever, in fact, been achieved. Instead, it is a condition of a complete description, not a claim to possess one.[453] Since such conceptual mastery demands mastery of a thing and its relations to everything else, the task of concept mastery is limitless and all knowledge gathering activities are, at best, asymptotically progressive. And yet there is one being, God, whose essence is necessary and whose knowledge is essential, and so who exhaustively comprehends all things as it knows itself. Ultimately, therefore, since there is a *single Whence* of all explanations, and that explanation (including its relations to everything else) is strictly necessary, there is no analytic/synthetic divide.

If all things are ultimately explained by one thing, and proximately explained by a single unified system, there is also no mind-matter dualism. Such a conclusion is hardly surprising given Schleiermacher's heavy use of Spinoza since Spinoza explicitly intended to re-integrate the world of thought and extension.[454] And although Schleiermacher does not share the Cartesian attributes of "thought" and "extension" with Spinoza, he does share the same *in principle* commitment to the ultimate unity of the mental and physical. This commitment is most clearly seen in Schleiermacher's own treatment of the divine attributes,

---

452 The means of discovery, i.e. whether something is discovered empirically or otherwise, do not determine whether a truth is analytic or not. A judgment is analytic when it follows, of necessity, from the terms adequately comprehended—that is, when its contrary implies a contradiction. The implicit content of notions can be discovered over time by experience. For instance, that Julius Caesar crossed the Rubicon had to be discovered by experience. Yet, on the supposition of absolute necessity—which, as we saw in Chapter 5, Schleiermacher claims in no uncertain terms—whatever is true of Julius Caesar is certainly true of him, because nothing can happen in any other order or way than it, in fact, does without contradiction. Therefore, the notion of Julius Caesar who did not cross the Rubicon is a contradiction. Meaning, Julius Caesar's crossing of the Rubicon is both discovered empirically *and* follows analytically from the very notion of Julius Caesar exhaustively comprehended. This is a natural extension of Schleiermacher's adoption of Spinoza's geometrical necessity and Leibniz's conceptual containment, since this is exactly what both notions explicitly or implicitly entail. For a more in-depth explanation of how this doctrine is entailed by absolute necessity, and for Schleiermacher's identification of metaphysical necessity with logical necessity, see Chapter 5.

453 "Both kinds of method will renew themselves side by side in diverse products until the highest knowledge and all the especial sciences are perfected simultaneously. This perfection is a goal, it is true, which can never be achieved outright, but an approximation to it should be seen in the fact that in each science, even in terms of content, the two kinds of method will coincide more and more closely; moreover the diverse shapes taken by each science as a result of one or the other will come closer together." Schleiermacher, *Lectures on Philosophical Ethics*, Introduction, final version, §17; 140.

454 Spinoza, *Ethics*, Vpref.

where omniscience receives its place precisely as the attribute of the ultimate explanation which signifies that that *Whence* upon which we are absolutely dependent is not less than thinking and living.[455] Even more importantly for our present purposes, the unity of thinking and being means that ideas are not exempt from causal explanations, and, like all explanations, terminate in absolute necessity *in principle*. In other words, through the ultimate unity of explanations, Schleiermacher is committed to something like Spinoza's doctrine that "The order and connection of ideas is the same as the order and connection of things."[456] Both things and ideas are aspects of one and the same nature system.[457] Therefore, thoughts, ideas, volitions—in short, whatever mental states, acts, or events can be conceived—are, in principle, included in the universal causal nexus which must be explained through a determinate causal chain terminating in a single self-explaining explanation, and which are not adequately comprehended until exhibited as necessary.

However, although Spinoza is happy to admit thought as an attribute of ultimate reality, his position on final causes is more ambiguous. He appears to attribute teleological action to humans while denying it to God.[458] Leibniz, however, has no such trepidation. And, taking care to qualify divine teleological action by denying imperfection (basically, by granting Spinoza all of his points), Schleiermacher, following Leibniz, is not only happy to admit final causes, but insists on them as requisite for a complete account of God and world.

Final causes are, like all causes, explanations. And, by the principle of the unity of explanations, they too must terminate in a single self-explaining explanation. But of course, Schleiermacher has also admitted and used other causal categories: efficient causes, and also important (though sometimes tacit) appeals to formal causes. But these causes must, in principle, terminate in one and the same self-explaining explanation as well. And that means that not only will one and the same explanation explain efficient *and* final causes (and formal causes, etc.), but that since a sufficient explanation is a single complete explanation, it is only through an abstraction that efficient and final causes can be divided. In reality, not only is the order and connection of ideas the same as the order and connection of things, but efficient and final causes are also different facets, or modes of consideration, of one and the same order. That means that not only will complete causal accounts of free actions be possible in principle, but that relatively free agents will not be outside the natural order. The actions of free fi-

455 *GL* §55; *The Christian Faith*, 219–22.
456 Spinoza, *Ethics*, IIP7.
457 *GL* §§47.1, 49; *The Christian Faith*, 178–81, 189–93.
458 Spinoza, *Ethics*, IApp.

nite agents are thus both *natural* and as exhaustively determined as any other occurrence in the nature system. Because of the unity of ultimate explanations, free finite agents and their actions do not transcend the natural order but are part of it.

These in-principle explanatory demands set the necessary conditions of any explanation, but they are not sufficient. Ultimately, the unity of explanations requires the unity of the necessity of the natural order, and the necessity of the goodness of divine purpose, *and the determinate content of both,* in a complete account. And Schleiermacher's defenders have for some time insisted that Christological content does exactly this in his theology. Importantly though, Christology does not modify principles. Alone, it only determines, and so perfects, content. It specifies the *Whence* that we already knew was necessarily self-subsistent, as love. Likewise, the investigation of the natural world does not modify principles, but it too only specifies content. It does not arrive at its highest principles from the investigation of the world but proceeds from them. Consequently, it cannot claim to exhaustively comprehend a world which it has not yet exhibited in its absolute internal necessity. The divine wisdom, which completes Schleiermacher's account of the world as good, requires *both* explanatory principles *and* determinate content. Neither is a sufficient condition on its own. Only when the world is known as a world and so known as ordered, *and* specified as ordered to redemption, is its order rendered wise and so the world known as good.

In the end, the same in-principle explanatory demands are preconditions for understanding both God and the nature system. In consequence, both efforts rise and fall together. That the same principles fund understanding of both God and world means that the benefits of either are only to be secured together. Should Christians reject the explanatory basis of natural science, they risk undermining their own basis of understanding God. And should natural scientists reject causal necessity, final causes, or the necessary terminus of explanation, they risk undermining their own enterprise by revealing it as incomplete or arbitrary. Without a complete causal account, including efficient and final causes, no sufficient account of the world can be given. Neither can a sufficient account of the world be given without reference to the *Whence* upon which it depends. When a sufficient account is given, it will turn out to be a complete account. Everything will be and be conceived through another or through itself, such that if anything were to be other than it is, it would be a contradiction. In a complete account, everything follows of necessity from the First Principle.[459]

---

459 "If any one especial science is to be perfectly depicted, it cannot begin purely on its own

Far from undermining theology or natural science, this has, unknowingly, been the intent of both all along when they attempt to attain to the quality of knowledge known to the ancients as *science.* Knowledge has reached the level of science when its conclusions are deductive, when all content follows, of necessity, from terms adequately comprehended. What Schleiermacher has done is simply to give each their wish. If there is to be science at all, natural or divine, it must, at least in theory, rise to the exhibition of the necessity of things.[460] But, if the necessity of any one thing is to be exhibited, it must include reference to all else: all finite things which relatively condition every other thing, and God who absolutely conditions all things. Therefore, in the final analysis, *the conditions of the perfection of natural science and divine science are one and the same.*[461] Not only do they rely on shared principles, but they also ultimately rely on shared content. They are two indispensable facets of a complete account. Without one another, neither the world nor God can be adequately comprehended. In fact, in the end they are not two utterly different activities, but merely two aspects of one activity. Natural science *is* theology indirectly insofar as it describes the divine essence under the aspect of divine means. And theology *is* natural science indirectly insofar as it names the love to which the world is ordered and the goodness without which it cannot truly be comprehended.

Delineation and Demarcation

Nevertheless, Schleiermacher does distinguish disciplines across a range of works, including *The Christian Faith.* How do I claim to reconcile these clear distinctions with my claim that theology and natural science are, on Schleiermacher's account, ultimately about the same thing? The answer, explained below, is that Schleiermacher distinguishes natural science and theology (and every other discipline) *only relatively.* In their *perfection*, which is to say in their absolute completion, they are identical. This answer is intimately related to the question of demarcation that I took up in the introduction. If Schleiermacher is committed to the *ultimate* demarcation of disciplines, as opposed to merely relative distinc-

but must be related to a higher knowledge, and finally to the highest knowledge, from which all individual knowledge must proceed." Schleiermacher, *Lectures on Philosophical Ethics*, Introduction, final version, §1; 136.

460 "Unless we regard them as being derived from the highest knowledge, all especial [branches of] science are simply the product of opinion." Schleiermacher, *Lectures on Philosophical Ethics*, Introduction, final version, §3; 137.

461 Once again, here I use these terms in their ordinary senses.

tions between them, then he is committed to an idea which has long been exploded in the philosophy of science and which has disastrous theological consequences.

Two disciplines can be distinguished absolutely or relatively. For example, one might claim that natural science has nothing to do with the good and its pursuit in human action, that they are absolutely distinct. On this account, they are two mutually exclusive topics which share nothing in common. They are ultimately *about* two entirely different things. Alternatively, one might claim that natural science and the study of human action are only relatively distinct: that they are, in important senses, about two different things, but not *ultimately* about two different things. When pressed, their boundaries are fluid, not fixed; and not only in the sense that their boundaries are contested, but that there is inevitable genuine overlap in their content. I do not contest that Schleiermacher distinguishes disciplines relatively. I only deny that he distinguishes them absolutely or ultimately.

What might the relative distinction of disciplines look like concretely? An ordinary example is the distinction we make between physiology and psychology in regard to the study of human persons. The discipline of physiology is not exactly the same as psychology. They employ different methods, they study distinct things. However, while physiology and psychology are distinct disciplines, they are not ultimately about two different things. In fact, they are explicitly about a single thing: namely, human persons. Further, the study of the human body inevitably overlaps with the study of the human mind, and vice versa. In fact, in order to perfectly understand the human body, one *must* understand human minds; likewise, in order to perfectly understand human minds, one *must* understand human bodies. The perfection of either includes the perfection of both. And so, in their perfection they are identical. That is, when complete, they are *about* one and the same thing. The alternative is mind-body dualism. While this is a genuine metaphysical option, it famously comes with its own burdens, the foremost of which is the problem of how to coordinate a mind and a body which can, in principle, have nothing to do with one another.

To delineate disciplines at the largest scale *ultimately*, as opposed to relatively, requires one to carve up reality as a whole in just the same way as mind-body dualism requires one to carve up human persons. Likewise, one faces just the same burden of answering how, exactly, two facets of reality relate to one another, even though they ostensibly have no relation in a higher unity. However, unlike mind-body dualism, at this highest level any effort to unite these disparate facets in a higher synthesis fails. On the supposition that two disciplines ultimately have nothing to do with one another, the two disciplines at hand can't even be about one common reality because then they would, in fact, ultimately be about one and the same thing, which is a contradiction. Therefore, either two disciplines are about one reality, in which case they are not ultimately different, or they are ultimately different, in which case they are not about one reality. But Schleiermacher is plainly committed to the unity of reality, as is clear both from his strong doctrine of a world and by his insistent use of the "principle of monotheism," as we saw in Chapters 4, 5, and 6. In some of Schleiermacher's works this highest unity is called "the Absolute," but in *The Christian Faith* it goes by the name of "God." To deny that all disciplines are ultimately about the same reality is to deny that there is ultimately *one* reality they can be about. God, as Schleiermacher makes clear, is the one ultimate reality through which all exists and can be explained. Therefore, to insist on the ultimate demarcation of disciplines is to deny that there is *one* God, one *Whence* upon which everything depends, one explanation which explains itself and all else. The consequence of *ultimately* demarcating disciplines is, in short, either polytheism or atheism.

One of the most popular approaches to demarcating disciplines over the last two-hundred years is the division between fact and value. As I discussed in the introduction, this well-known idea was reinvigorated in late twentieth century debates about religion and science by Steven Jay Gould with his account of non-overlapping magisteria, though its genesis lies in Kant's critical philosophy. Science, on this view, is about facts. Religion is about values. Since facts and values have nothing to do with each other, they cannot conflict. The utter division between the two keeps the peace. For these reasons this approach carries appeal. Yet, it too suffers from all the same problems as any ultimate demarcation, and is thus equally wanting. However, beyond the incompatibility with Schleiermacher's claims and practices, and beyond the savaging the fact-value divide has received in the philosophy of science, there is a more pressing theological problem with the fact-value divide that brings us back to the unity of explanations.

To claim that distinctions between disciplines are more than merely relative is, ultimately, to deny that there is a single ultimate reality. That is, demarcation contradicts monotheism by denying the notion of a unique self-explaining explainer. To divide facts and values is no less theologically problematic. For, as with *any* ultimate division, it will result in a plural ultimate reality. In this case, fact, that is, being, is divided from value, that is, goodness. Since, on the supposition of their division, they cannot be one. So either ultimate reality cannot be good, or else the ultimately good cannot be real. Moreover, as we saw in Chapter 4, Schleiermacher equates the being of a thing with its power. That is, to exist *is* to be a locus of intrinsic forces and powers. That means that a thing that was ultimately real is ultimately powerful. But, by the fact-value divide, this being/power cannot be ultimately identical with value/goodness. Therefore, the fact-value divide amounts to dividing God's being and goodness, that is, to denying that either God's omnipotence or God's omnibenevolence is properly essential. Whatever God is, God cannot both *be* and be *good* equally, naturally, intrinsically, and at once. On the supposition of the fact-value divide, the divine nature can only either be a valueless omnipotence—naked power, unadorned by its own excellence, undesirable by any creature, unloving and unwise—or, God must be an impotent goodness which, because of its total lack of power, cannot really *be*, and so must be a mere imagination. The fact-value divide, even if it does claim to admit of a single so-called "God," nevertheless denies the basic Christian equation of God's being and goodness, power and value, and certainly contradicts Schleiermacher's clear insistence that the only *truly* adequate description of God is as "omnipotent love."

No, Schleiermacher does not ultimately partition reality, as one feature of reality from another, or as fact from value. Instead, Schleiermacher holds to the ultimate unity of reality as well as to the ultimate unity of knowledge. And this is exactly what we should expect, no less of Schleiermacher the philosopher than of Schleiermacher the theologian: both in the theological tradition broadly, and in Schleiermacher's own work, the unity of knowledge and the unity of ultimate reality are corollaries. As God is one, so is truth.

Because, on Schleiermacher's account, disciplines cannot be divided ultimately without partitioning the world and God, they must, in the end, be *about* the same thing, however indirectly. That is, the study of anything, under any aspect or mode, must eventually bear on the study of everything else. Moreover, because this is true of all disciplines, all topics, all facets of reality, no knowledge will be perfect and complete unless it is connected to everything else and ultimately to God. It is no surprise then that we find Schleiermacher insisting over and over across a range of works that, although disciplines can and should be distinguished, they are all ultimately interconnected and, even *require* one another for their mutual perfection.[462] This is what these texts make clear. This is what Schleiermacher's doctrine of God in *The Christian Faith* entails. And this is precisely the kind of cooperative union that one would expect of a covenant.

However, two disciplines which ultimately share the same object do not necessarily share the same *proximate* aim or task. Because of this, nothing about the unity of knowledge contradicts Schleiermacher's claims in *The Christian Faith* or

---

462 See, for example, Schleiermacher, *GL* §17; *The Christian Faith*, 83–84; *Brief Outline in the Study of Theology*, trans. Terrance Tice (Louisville: Westminster John Knox, [1811, 1830] 2011), §16; *Introductions to the Dialogues of Plato*, 388; Schleiermacher, *Lectures on Philosophical Ethics*, Introduction, final version, §§ 6, 61.

in the *Brief Outline of Theology as a Field of Study* about the role of dogmatics vis-à-vis the church. Dogmatics is, on Schleiermacher's account, *for* the church. It seeks practical ends: the proximate end of serving as an aid to preaching and the distant end of propagating and clarifying the God-consciousness of Jesus of Nazareth.[463] Natural science does not share those practical ends, at least not proximately—although as we saw above, the consciousness of God and world are *ultimately* mutually perfecting. Like physiology and psychology, they can be studied for different proximate ends. For example, one might study physiology to prevent illness, but study psychology to discover how humans learn best, though both ultimately regard the perfection of human being. In the same way that disciplines can be relatively distinguished according to object, they can also be relatively distinguished according to aims and use. The distinction of disciplines according to proximate purposes in no way contradicts their ultimate unity.

In addition, two disciplines might be relatively distinguished according to their main standards of adequacy, that is, according to the phenomena they are tasked with explaining. For example, physiology has as its primary phenomena-to-be-explained the structure of the human body: bones, ligaments, muscles, and organs. It seeks to explain these insofar as they appear to function individually and as a body. A physiologist might seek to explain how a person runs or why they can't, but rarely do physiologists, as physiologists, seek to explain how a person does their taxes or why they won't. They typically restrict themselves to explaining bodies as bodies in accordance with their aims and their relative matter-about-which. Likewise, psychologists know that their patients need bodies to talk to them and to act, but the phenomena they explain are thoughts and behaviors, not bodies as bones and ligaments. They have as their relative matter-about-which human beings as agents and so they prioritize some phenomena over others. In this way, the phenomena-to-be-explained of two disciplines can be relatively distinct without contradicting the ultimate unity of their objects.

Dogmatics, for Schleiermacher, works the same way. The phenomena to which Christian theologians are accountable is, first and foremost, the Christian consciousness of redemption in Christ as an *effect* on consciousness. Secondarily, this consciousness is reflected in scripture, creedal confessions, the writings of the Church Fathers, and in the history of the reflection on doctrine itself. Consciousness of redemption is the principle standard of adequacy for Christian re-

463 Schleiermacher, *Brief Outline of Theology as a Field of Study*, §§196–202; *GL* §15–18; *The Christian Faith*, 76–88.

flection on God. It is the phenomena dogmatics seeks to save. However, dogmatics is not *about* that consciousness.[464] Instead, dogmatics has God for its matter-about-which, its object. Dogmatics is *adequate* to the Christian consciousness of God, *for* the church, but is *about* God.

Any number of disciplines can be relatively distinguished according to these or any other means, provided that their distinction is not taken to denote two or more *ultimate* matters-about-which that are, in the final analysis, really different. Schleiermacher's distinction between disciplines according to aims, objects, and standards of adequacy likewise does not commit him to ultimately demarcating disciplines. Since truth is not ultimately plural, it is ultimately one.

The Eternal Covenant and Its Implications

The eternal covenant has been misunderstood. It is not a proposal to partition disciplines to keep the peace, nor is it a proposal to resolve conflict or disagreement on the basis of a pre-arranged disciplinary hierarchy. Instead, it amounts to a cooperative union, a friendship, between theology and science in pursuit of a shared perfection.

The basis of the eternal covenant has also been misunderstood. The eternal covenant is not a mere methodological proposal. Instead, it follows from a series of arguments. These arguments begin with at least some publically shared principles and assumptions—especially the notion of a world and the principle of sufficient reason. However, Schleiermacher also leverages specifically Christian commitments supplied by the feeling of absolute dependence as determined by the redemption accomplished by Christ, such as the consciousness of being absolutely dependent on a single *Whence* as well as the teleological order of the world.

These commitments have content: the consciousness of being absolutely dependent on a single *Whence* implies that all depends on one highest being and it on nothing; the Christian consciousness of redemption implies that this *Whence* remains merely an indeterminate something until its actions are specified as wisely ordered to the end of divine love. Schleiermacher's case for the eternal covenant amounts to showing that, although the Christian consciousness of redemption can be shared by all Christians equally, there is only one way of *explaining* that consciousness that does not yield contradictions upon thorough examination: namely, that God acts for the one end of love, in perfect wisdom, by

464 *GL* §50; *The Christian Faith*, 198–199.

natural necessity. It is that account, in turn, that implies that God never acts except through secondary causes, and yet which simultaneously implies that the world is, as a unity of plans and purposes, means and ends, the perfect artwork of God in which the character of God finds complete expression, and is therefore the absolute revelation of the divine essence.

Because the basis of the eternal covenant is this specific and complex account of God and divine action, any reading of Schleiermacher on method must cohere with, and warrant, these claims. That is, not only must any reading of Schleiermacher's method be consonant with this account, it must be able to explain Schleiermacher's account in detail and, at the same time, must be able to explain Schleiermacher's claims as moves in an argument intended to expose the inconsistences and shortcomings in other accounts of God, the world, and divine action.[465] In short, any reading of Schleiermacher on method must not only explain the *form* of Schleiermacher's case but be able to organically relate that form to its concrete *content*, and to Schleiermacher's persuasive *ends*. This approach differs from the long tradition of reading Schleiermacher that assumes that his thought can only be properly understood by *beginning* with his remarks on method. The truth is actually the reverse: Schleiermacher's remarks on method can only be properly understood in light of his clear, concrete claims supported by detailed arguments as we saw in regard to his accounts of natural development, his account of a world and absolute versus relative miracles, his criticisms of absolute miracles on the basis of divine perfection, and his extraordinary account of the world as the revelation of God.

This reading of Schleiermacher matters for several reasons. First, this account makes many neglected arguments and claims in *The Christian Faith* (on God, world, miracles, etc.) fully intelligible. Other interpretations present Schleiermacher's claims, but not the theological or philosophical reasons for these claims in detail. In contrast, the account I offer explains *exactly* why Schleiermacher's conclusions follow from his premises, and even why those premises might command assent. Schleiermacher's account of divine action follows from the perfection of God, the uniqueness of God, the creator-creature distinction, and more. At minimum, alternative readings must be able to explain these claims and arguments in equal rigor and detail.

Second, this reading matters because, once understood in detail, the content of these claims shows itself to be *concrete*, that is, to be determinate claims organically related to general principles as a unity of form and content. The concrete content of Schleiermacher's claims in turn reveals the error of his critics.

---

465 Whether Schleiermacher is judged successful or not is another matter.

Whether in the nineteenth century or the twentieth, Schleiermacher's thought has been taken by his critics to be generic to a fault and/or premised on mistaken methodological presuppositions.[466] My account shows that charge to be false: not merely with regard to method or intent, but with regard to means and execution as well. Instead, throughout *The Christian Faith* Schleiermacher's claims are neither generic nor based on mere methodological decisions, but are specific claims that are, Schleiermacher argues, entailed by shared premises such as the principle of monotheism, the principle of sufficient reason, and so on.

Third, since, as I demonstrate, Schleiermacher gave powerful arguments for these claims, his account of God is not merely one option among others, but a contender for the status of *best*. Schleiermacher meant his account of God, the world, and their relation to be judged better than its alternatives, to persuade us to discard other views in favor of his. As his arguments show, Schleiermacher's position deserves to be taken very seriously indeed. His arguments are detailed, powerful, plausible, and so not only meant to be persuasive, but, hopefully, actually so. Of course, Schleiermacher's positions are extremely controversial. Schleiermacher recognized this, and knew that if he was to persuade those who disagreed with him, he would need to appeal to assumptions shared in common with his opponents and then to show how those assumptions led to his conclusions, not theirs. Schleiermacher seeks to make a compelling case for any who subscribe to a largely traditional view of God to see that that tradition reduces to his account of God, and then to see that *that* doctrine of God in turn implies a specific view of divine action, a view which then implies a specific view of the natural order. Hence, any wishing to resist his view of the natural world will have to face difficult consequences in their doctrine of God.

Arguments, Ideas, and Sources

Historically, the conceptual analysis I offer bears most heavily on evaluating Schleiermacher's influences. The account I advance makes Schleiermacher's eternal covenant fully intelligible in detail by showing how important concrete accounts of God, divine action, and the natural order were to his theology and how important particular arguments were to these accounts. In consequence,

---

466 See, for example, Emil Brunner, *Die Mystik und das Wort* (Tubingen: J.C.B. Mohr [Paul Siebeck] 1924); Karl Barth, *The Theology of Schleiermacher*, trans. Geoffrey Bromily (Grand Rapids, MI: Eerdmans, [1923/24, 1978] 1982); Wayne Proudfoot, *Religious Experience* (Berkeley: University of California Press, 1985).

these arguments—their sources, content, and presuppositions—in turn all count against certain evaluations of Schleiermacher's influences.

Above all, the account I advance qualifies the influence of Kant's critical philosophy on *The Christian Faith.* In fact, on a number of important points, Schleiermacher's claims in *The Christian Faith* are incompatible with Kant's *in principle.* This does not mean Kant had no influence. That would be an exaggeration. Yet, my analysis shows that the view that the eternal covenant, in particular, is based *primarily* on Kantian grounds is false. We discover this precisely by turning from methodological issues to Schleiermacher's concrete claims about God, divine action, and the natural world. Those claims have content and are supported by arguments, all of which have historical sources. Not only is Kant's critical philosophy[467] not a source for those specific claims, but the figures who are sources—Leibniz and Spinoza—are *alternatives* to Kant in important respects, and their views, the very claims and arguments Schleiermacher takes up, likewise explicitly contradict crucial Kantian tenants. The conclusion I argue for is, however, not a total dismissal of Kant, but rather the careful and specific qualification of Kant's influence on these important points.

Most importantly, Schleiermacher's position contradicts Kant's in affirming the ultimate unity of knowledge. This is a denial both of Kant's fact-value divide as well as the related noumenal-phenomenal divide. On Kant's view, the two sides of these respective divides are mutually exclusive. On Schleiermacher's account, in contrast, they are mutually perfecting.

This departure follows from Schleiermacher's denial of Kant's analytic-synthetic distinction. This distinction is crucial because without a distinction between analytic and synthetic judgments, Kant cannot sustain his category of synthetic *a priori* judgments. To deny this prior distinction is, then, to deny the basis for Kant's critical philosophy at large.[468] Schleiermacher does just this when he takes up Leibniz's and Spinoza's accounts of necessity. This holds even in regard to Leibniz's distinction between truths of reason and truths of fact since, on Leibniz's account, truths of fact reduce to truths of reason in the final analysis—as his doctrine of conceptual containment implies. Spinoza is even clearer. Everything happens as it does of absolute necessity. If the world were other than it is, down to the smallest detail, it would imply a contradiction. Since in *The Christian Faith* Schleiermacher adopts Spinoza's geometrical (i.e. absolute) necessity

---

467 The *Universal Natural History and Theory of the Heavens* was written well before Kant's development of the critical philosophy for which he is known and does not rely on, or imply, his later work.

468 As Frank says, regarding Schleiermacher's account of judgment, "This view is thoroughly un-Kantian." Frank, "Metaphysical Foundations," 22.

in no uncertain terms, he correspondingly denies the analytic-synthetic distinction, since everything follows as it does from the divine essence such that if it were other than it is in any way whatsoever, it would be a violation of a geometrical, i.e. analytic, truth. In short, when Schleiermacher adopts Leibniz's, and, especially, Spinoza's views on truth and necessity, he *necessarily* rejects Kant on these matters.

This does not mean that these necessary truths aren't discovered over time, by experience, as a result of empirical study. They are. Both Spinoza and Leibniz held this view as well, at the same time that they held that all truth was contained in the notions of things adequately comprehended. No, the status of a truth as analytic, or otherwise, only regards the very basic question—however we come to discover a truth—of whether its contrary implies a contradiction.[469] Schleiermacher, following Spinoza and Leibniz, indeed thinks just this, both according to Leibniz's necessity of the best (see Chapter 3) and according to Spinoza's absolute necessity (see Chapter 5). Since nothing about the *means* of discovery or the *order* of discovery contradicts the claim that all true judgments follow from their concepts adequately comprehended, nothing about Schleiermacher's commitment to the empirical discovery of knowledge stands in the way of understanding how, for Schleiermacher, all truth is ultimately analytic as following of absolute necessity from the notion of God adequately comprehended.

But that is not all. Schleiermacher also rejects another distinction, which is the subject of another of Kant's works, the *Critique of Judgment*.[470] There, Kant distinguishes between the sublime, the beautiful, and the purposive. And he distinguishes between these three in such a way that they are *mutually exclusive*.[471] The sublime is not the beautiful which is not the purposive. Schleiermacher re-

---

469 "In all judgements in which a relation between subject and predicate is thought [...], this relation is possible in two ways. Either the predicate B belongs to the subject A as something which is (covertly) contained in the concept A; or B lies outside the concept A, though connected with it. In the former case I call the judgment *analytic*, in the latter *synthetic*." And further: "For *if a judgment is analytic*, whether negative or affirmative, its truth can always be sufficiently known according to the principle of contradiction. For the opposite of that which already lies, and is thought as a concept, in our knowledge of an object is always rightly denied; but the concept itself must necessarily be affirmed in it, for the simple reason that its opposite would contradict the object." Immanuel Kant, *Critique of Pure Reason*, trans. and ed. Marcus Weilgelt, trans. Max Müller (London: Penguin, [1781, 1787] 2007), B11, B191.

470 For a detailed comparison of Kant and Schleiermacher on a number of relevant points see Anne Käfer, "Kant, Schleiermacher und die Welt als Kunstwerk Gottes," *Zeitschrift für Theologie und Kirche*, 101 (2004), 19–50.

471 Immanuel Kant, *Critique of Judgement*, trans. Werner S. Pluhar (Indianapolis: Hackett, [1790] 1987), §§11, 15, 23.

jects this exclusivity. In fact, as we saw in regard to the world as divine artistry, these three are *mutually perfecting*. We are conscious of being absolutely dependent on a *Whence* of our being, the absolutely sublime. But this sublimity is only rendered determinate—that is, perfect and complete—when its action is specified as ordered to love, that is, when it is reckoned as a *purposive* sublimity, a *Whence* whose determinate purpose is to communicate its goodness. Further, the unity of wisdom and love, means and ends, makes the world the perfect undivided appearance of its sublime reality. The world is an artwork of God which perfectly reveals the character of its maker. And as a perfect work of art, it is supremely beautiful. Therefore, the one *Whence* upon which we depend absolutely is only incompletely known as that alone. In its perfection it is the *identity* of the sublime, the purposive, and the beautiful in one: a single almighty-wise-love which cannot be truly known unless it is known as indivisibly so. And, unlike for Kant, what pleasure we derive from confrontation with the sublime is not due to the power of our minds in overcoming it, the thrill one feels when one safely flirts with danger.[472] No, for Schleiermacher, the pleasure of redemption is not in overcoming the power of the sublime with our minds, but in our minds receiving, by the power of the sublime, the almighty-wise-love communicated through the redemption accomplished by Jesus of Nazareth.

Finally, for Schleiermacher there is no noumenal-phenomenal divide. True, on Schleiermacher's account, we cannot have comprehensive knowledge of God in the strictest sense. But this is not at all for Kant's reasons (see below). In the first place, Schleiermacher has rejected the very basis of Kant's noumenal-phenomenal divide when he affirmed the absolute necessity of all things as entailments of the divine essence. In order to affirm Kant's conclusions for Kant's reasons, Schleiermacher would have to recant his doctrine of God.

In the second place, there is simply no recourse to Kant needed to explain Schleiermacher's perfectly traditional view—a view which is explicitly affirmed by Schleiermacher for traditional, not Kantian, reasons. I explain this claim in detail elsewhere,[473] but, in short, Schleiermacher's position follows from a consideration of the infinity, simplicity, and uniqueness of God, as well as from a classical understanding of *knowledge*. Knowledge, in the strong sense, can properly be had only when the object of knowledge is *defined*. In order to be defined, said object must be comprehended under a higher principle. In the case of God, it is impossible to define God under a higher principle because *there is no higher*

---

472 Kant, *Critique of Judgement*, §28.

473 Daniel Pedersen, "Schleiermacher and Reformed Scholastics on the Divine Attributes," *International Journal of Systematic Theology*, 17 (2015), 413–431.

principle. God is not a member of a genus, not a thing among things, and so God cannot be defined.[474] Because God cannot be defined, i.e. comprehended under a higher principle, God cannot be comprehensively *known* in this strong sense.

However, while God cannot be defined, on Schleiermacher's account, God can be described. That is, something true can be said about God as the sufficient reason of God's effects that gives us good reason to make claims *about* God which are perfectly adequate for our own purposes, but necessarily inadequate to the divine essence they attempt to describe. Why? By virtue of God's infinity, no finite descriptions can ever hope to approach the divine essence in scope. And, because any finite descriptions of God must necessarily be compound, they are necessarily inadequate to the utterly simple divine essence which they are about. Nevertheless, we can describe God, albeit incompletely, by attributing God's effects to God as cause, which is to say, as that which contains within itself the sufficient reason why its effects come to be. And this is exactly what Schleiermacher does in his dogmatics. These effects of God are effects on our consciousness. In giving an account of God, these effects on our consciousness are attributable to God as cause, that is, as the sufficient reason for these effects.[475]

Schleiermacher's modesty about knowledge of God can be explained without recourse to anything but the theological tradition running from the early Church through Augustine, Anselm, Thomas, and the Reformed tradition of which Schleiermacher was a part. And, if we did compare this account to Kant's, we would find important differences. For Kant, God is only a practical postulate. For Schleiermacher, God is the *Whence* which we are conscious of being absolutely dependent upon. For Kant, cause and effect relationships are synthetic *a priori* impositions of consciousness. For Schleiermacher, causes are the genuine sufficient reasons why their effects come to be. No, Schleiermacher's view comports much better with the tradition traced back through both Reformed Scholastics and thinkers like Thomas Aquinas, than it does with Kant. And because it comports better with the tradition than with Kant, we needn't have recourse to Kant to explain Schleiermacher's position on the impossibility of knowledge of God, in this strict sense.

These important differences offer substantial reason to reconsider how strong Kant's influence was in formulating the basis of the eternal covenant. Surely Kant cannot be entirely dismissed, but neither did Schleiermacher sub-

---

474 *GL* §§4.4, 10 postscript, 50; *The Christian Faith*, 18, 52, 194–199.
475 *GL* §50.3; *The Christian Faith*, 198.

scribe to Kant's project at large. As my analysis shows, there are *central* disagreements between Schleiermacher and Kant that cannot be reconciled.

Instead, as I have hinted throughout this work, I take Schleiermacher's views to be best explained as rooted in a tradition looking back to Plato. This hypothesis, which can only be fully substantiated in future projects, embraces all of the major claims in *The Christian Faith* revolving around the doctrine of God, divine action, the natural order, and Schleiermacher's corresponding account of the unity of knowledge. As Plato teaches, the lower sciences are rendered intelligible by higher principles, either proceeding from a first principle, or as provisionally assumed by the lower sciences themselves for the purposes of rendering the phenomena of the sensible world reasonable. This is exactly what we saw natural scientists themselves doing in Chapter 2. But Plato does not stop there. Dialectic proper on his account, "treats assumptions not as principles, but as assumptions in the true sense, that is, as starting points and steps in the ascent to something which involves no assumption and is the first principle of everything; when it has grasped that principle it can again descend, by keeping to the consequences that follow from it, to a conclusion."[476] In this way, the phenomena, mediating principles of lower sciences, and, finally, first things are united in a common complete account of everything that, ultimately, follows of necessity from the First Principle but which must be discovered over time by experience. In other words, Plato subscribes to the unity of knowledge, and the relationship between the First Principle, i.e. God, and the lower sciences is just as it is for Schleiermacher, which is to say, as *theoretically* (though not practically) deductive.

Moreover, in Plato's story of creation in the *Timaeus*, we see this general process yielding an account full of details about the particulars of divine action and the natural world. God does everything only for the best of reasons, that is, for the sake of the good.[477] And because the world God makes perfectly serves God's good ends, there is no reason for God to act other than through secondary causes.[478] This is what we saw Schleiermacher argue for in Chapter 3. In turn, the world is the perfect image of its invisible maker, a whole which unites efficient and final causes seamlessly into a perfect and unified whole. This world is no machine, but as a natural unity of efficient and final causes is, instead, an organism.[479] Because God cannot create any world but the best, and because the best alone is the reflection of the supreme goodness that made it, the world is there-

476 Plato, *The Republic*, trans. Desmond Lee (London: Penguin, [1987] 2007), IV, 6; 239.
477 Plato, *Timaeus*, 29e.
478 Plato, *Timaeus*, 46c8-d1.
479 Plato, *Timaeus*, 29b7-c1.

fore unique.[480] The above claims are what we saw Schleiermacher argue for in Chapter 4. And because it is the nature of goodness-itself to be good, God creates by the necessity of the divine nature.[481] This is what Schleiermacher argues for when he secures the necessity of divine action, as we saw in Chapter 5, as well as when he unifies the divine necessity with the divine wisdom and love, as we saw in Chapter 6. Finally, the world that God creates is a *perfect* artwork in exact antithesis to the fraudulent art Plato denounces.[482] The artwork that is the world is not made for profit or deceit.[483] It is, instead, a creation which reflects the artists' perfection, exhibited in the art itself, of representation, manufacture, and use.

The account of the eternal covenant I advance suggests Schleiermacher synthesized diverse arguments from Leibniz and Spinoza in service of his own distinctly modern theology anchored in Plato's basic insights. The detailed arguments Schleiermacher gives regarding God, divine action, and the world support this reading well, while, at the same time, those same claims and arguments contradict important claims of Kant's.

The Eternal Covenant Today

The account of Schleiermacher's eternal covenant I advance does not only have historical import. It also promises to intervene in contemporary debates about divine freedom, divine action, and the God-world relation.

Any number of contemporary accounts of divine freedom hold that contingency is necessary to secure God's freedom. That is, many assume that divine freedom must be such that for God to have done otherwise, even if God cannot *now* do otherwise, it would not imply a contradiction. Recent debates have revolved around this central issue. And yet, the question of the necessary and sufficient conditions of divine freedom is not a new concern. Instead, conversations on this theme run from the early church through the middle ages to the present. Condemnations and accusations centered on it. The issue of divine freedom thus remains a likely locus of debate and disagreement for the indefinite future.

Schleiermacher, on my account, joins that conversation by giving a controversial and sophisticated argument for divine freedom to be seen as *perfectly* compatible with natural or absolute necessity. That argument can be engaged in other times and places and for other reasons than those of Schleiermacher's

480 Plato, *Timaeus*, 31a.
481 Plato, *Timaeus*, 29a.
482 Plato, *Timaeus*, 92c
483 Schleiermacher, *Introductions to the Dialogues of Plato*, 357.

own time and place. I myself hope Schleiermacher's argument is engaged in just this way: as a normative proposal for how best to think about divine freedom.

In particular, I intend my explanation of Schleiermacher's position to contend with an important and popular assumption in contemporary debates. That assumption in regard to divine freedom is that necessity is precisely that which threatens divine freedom, while contingency is that which secures it. Even those who press for necessity of a sort to secure God's faithfulness, or even immutability, often find themselves qualifying that very necessity of divine action so as to maintain God's freedom. The tacit premise here is clear: something is not free when it is necessary. And this is, indeed, an ancient and respectable notion of both freedom and necessity. If this is true, it seems absolute necessity and divine freedom are, in fact, incompatible.

However, Schleiermacher, following Spinoza, argues that it is actually absolute necessity which alone *preserves* divine freedom. Not only is this claim, which was developed in full in Chapter 5, an important challenge to contemporary debates, but Schleiermacher's account of God's willing for the sake of the divine goodness, and his fulsome distinction between God and world secure his account against many of the most important objections to God acting by absolute necessity. Schleiermacher, in other words, offers an account of divine freedom on entirely different terms than the prevailing consensus.

Schleiermacher's account of divine freedom seamlessly and simultaneously bears on his account of divine action. His account of divine action, too, is in stark contrast to the typical accounts on offer. These likewise begin with assumptions: namely, that if God acts in regard to the world it will take the form of particular divine actions *within* nature, directed here or there to some particular end. The motivation for this account is just as noble and pious as that behind worries about divine freedom. It is an attempt to secure God's reactive response to prayer, biblical miracles, and more.

But despite its good motives, this popular account of divine action is also based on problematic assumptions. The issue at the largest scale is the judgment that, in the final analysis, there are only two accounts of divine action—deism, or some kind of divine intervention—on the supposition that in order for God not to be absent from the world, God would have to act within or upon it in particular ways. Leaving aside the question of deism (which merely functions as a term of accusation, only clear enough for one to know that it is a bad thing but not so clear that one is accountable for its careful use), the issue is, at bottom, a mixture of two related questions: one, the relation between divine preservation, created being, and power; and two, whether to think of God as a univocal cause.

The belief that mere preservation alone implies that God is somehow distant to or absent from the world reveals the tacit assumption that, if God preserves the world, God preserves the world as bare being: being without power and action. For, if in preserving the world, God preserved the world as a locus of intrinsic forces and powers, then not only would God's preservation be immediate and enduring (i.e. the contrary of distant or absent), but created forces and powers would be precisely those secondary causes *through which* God acts as primary cause. The failure to grasp the primary cause/secondary cause distinction is yet more evident in the solution to the perceived problem: that, in order to interact with the world, God acts within or upon created things. In sum, when God does not act in particular ways, God does not act; and when God does act, it is always within and upon particulars things in the world. This mutual exclusivity implies that God is conceived as a univocal cause. And it is *this* assumption which not only generates the false dilemma between deism or particular divine acts, but also causes a host of other theological problems in *loci* ranging from the notion of a world, to creaturely freedom, to providence and evil.

Schleiermacher's account is an alternative to the reigning accounts of divine action, especially those produced by theology-and-science specialists. And yet it is also an alternative to traditional accounts of divine action which make the primary cause/secondary cause distinction but still uphold miracles as divine acts which go beyond not only the natures of created things, but nature itself. Crucially, this account is motivated by the same piety of its alternatives, only to opposite conclusions. In this way, Schleiermacher's account of divine action is a decidedly modern account that not only does not require special divine action, but that sees every particular divine act as a contradiction of either divine goodness or divine power, or both, because all such acts imply imperfection in the necessarily perfect work of the necessarily perfect being.

Finally, Schleiermacher treats these issues as intrinsically connected. One cannot give a full account of divine freedom without giving an account of divine action, and one cannot give a full account of divine action without giving an account of the world. Schleiermacher's account is not only one of the most compelling accounts of each of the above topics individually, but also—and this is critical—gives a uniquely powerful *unified* account of all together. Each doctrine implies the other. And so, in order to truly reckon with Schleiermacher's position, his critics will have to engage all of the above questions simultaneously. Schleiermacher's challenge not only promises to call those with differing views to offer careful arguments from closely examined assumptions, but also promises to elicit alternative accounts which are likewise systematically responsible.

However, a final objection stands against the persuasiveness of Schleiermacher's doctrines of God, world, and divine action as a viable contemporary account. Even if Schleiermacher's case is deemed persuasive in its own context, the reader might worry that things have changed too much since Schleiermacher's day for it to apply any longer, particularly because natural science has discovered that the world of quantum theory is, in fact, indeterminate. Since Schleiermacher's account of absolute necessity is as determinate as possible, there seems to be an impasse. Schleiermacher's account would seem to imply that we take the natural sciences seriously, and yet the findings of those very sciences might be thought to contradict his account. In this last section I dispel this worry and conclude that there is no reason to doubt that the eternal covenant is as viable as ever, in spite of objections from quantum indeterminacy.

First, quantum theory is not clearly and necessarily indeterministic. There are divergent interpretations of the phenomena, including those that explicitly advertise themselves as deterministic. Even those most often thought to imply indeterminacy do not necessarily do so.[484]

Second, and more importantly, even if quantum theory was necessarily indeterministic, in order for quantum indeterminism to bear on metaphysical determinism of the kind that Schleiermacher holds, a move from theory to reality is required. Further, any such move which restricts itself to quantum theory, which claims that quantum theory *alone* should be held realistically, will not suffice. It will ring false of the *ad hoc* and the authoritarian. What is needed is a general claim under which quantum theory can fall as an individual instance. In order to move from theory to *reality*, what is needed is a notion of *scientific realism* to act as a major premise. The argument would then proceed as follows:

1. The best scientific theories of the day (or similar criteria) tell us what the world is *really* like.
2. Quantum theory is one such theory.
3. Quantum theory is necessarily, or at least most plausibly, indeterminist, etc.

But it is precisely here where the argument becomes most controversial. It is far from obvious that scientific realism, in either its stronger or more modest forms, is right.[485] And anything less than scientific realism fails to satisfy the major premise, and so fails to give traction to the minor. All that is required to doubt the relevance of quantum theory to metaphysical determinism is substantial doubt thrown on the major premise, and it is difficult to think of a more controversial premise than scientific realism.

---

484 Nicholas Saunders, *Divine Action and Modern Science*, (Cambridge: Cambridge University Press, 2002), especially 127–172.

485 For a sample of classic criticisms and replies see Jarrett Leplin, ed. *Scientific Realism* (Berkeley: University of California Press, 1984).

Finally, and most importantly for Schleiermacher's case, the very notion of an indeterminate world is synonymous with an incomplete account. This is a not a rhetorical trick. Following Leibniz and Spinoza, Schleiermacher holds that a fully determinate account is one in which a sufficient reason can be given for everything in principle (though, again, not necessarily in practice). A sufficient reason is a complete reason. Nothing more or less is needed for the thing in question to be what it is and no other. An ontologically indeterminate (not merely epistemically indeterminate) account of the world is, in contrast, one in which, it is claimed, there actually *is no sufficient reason* for everything in principle. For Schleiermacher, as a subscriber to the principle of sufficient reason, this is an absurdity. A sufficient reason is required for everything in principle in a complete account; and because anything less than a sufficient reason is an insufficient reason, an indeterminate account is, *in principle*, an incomplete account.

This matters for Schleiermacher's case for two over-arching reasons. The first is that, as we saw in Chapters 3, 4, 5, and 6, his leveraging of the principle of sufficient reason was tethered to the basic Christian principle of monotheism: that everything is ultimately completely explained by one and only one self-explaining explainer. God, in other words, is the sufficient reason for everything. In order to deny Schleiermacher's determinism, one must deny the principle of sufficient reason, but in order to deny the principle of sufficient reason, one is committed to denying that there is one ultimate reason for everything. Schleiermacher's account presses upon the opponents of determinism that a corollary to that belief is monotheism. In order to deny one, they must be prepared to deny both.

The second reason the incompleteness of indeterminate accounts matters for Schleiermacher's case is that, by the unity of knowledge, any one discipline or facet of knowledge is *necessarily* incomplete in isolation. Should physicists decide, on the basis of physics alone, that the world was indeterminate, that is, an incomplete world, Schleiermacher would heartily agree. He would insist that that is exactly what one should expect because the highest knowledge is an "essential identity of ethics and natural philosophy."[486] Natural philosophy, that is, what we call natural science, is an indispensable but incomplete description of the world which can never, in principle, yield a complete account of things. The same is true of all other disciplines, including theology. Only as a mutually perfected whole, a unity not only of discipline with discipline, but also a unity of principles and content, can we hope to attain to a complete account of the world.

---

486 Schleiermacher, *Introductions to the Dialogues of Plato*, 413.

Bibliography

Adams, Norman. *Schleiermacher's Philosophy of Freedom and His Relation to the Leibnizian Tradition*. Dissertation. Syracuse University, 1941.
Adams, Robert Merrihew. *Leibniz: Determinist, Idealist, Theist*. Oxford: Oxford University Press, 1994.
–. "Malebranche's Causal Concepts." In *The Divine Order, the Human Order, and the Order of Nature*. Edited by Eric Watkins. Oxford: Oxford University Press, 2013.
–. "Schleiermacher on Evil." *Faith and Philosophy* 13, 4 (1996): 563–583.
Alexander, H. G., ed. *The Leibniz-Clarke Correspondence*. Translated by Samuel Clarke. Manchester: Manchester University Press, 1956.
Anselm of Canterbury. *Monologion*.
Augustine of Hippo. *De Trinitate*.
Barth, Karl. *The Theology of Schleiermacher*. Translated by Geoffrey Bromily. Grand Rapids, MI: Eerdmans, [1923/24, 1978] 1982.
Basil of Caesarea. *Hexaemeron*.
Boethius. *Consolation of Philosophy*. Translated by H. R. James. London: Eliot Stock, 1897.
Brandt, Richard. *The Philosophy of Schleiermacher*. New York: Harper, 1941.
Bruckeri, Iacobi [Jacob Brucker]. *Historia Critica Philosophiae*. Leipzig, 1744.
Brunner, Emil. *Die Mystik und das Wort*. Tubingen: J.C.B. Mohr [Paul Siebeck]: 1924.
Carnap, Rudolf. "Ueberwindung der Metaphysik durch logische Analyse der Sprache." ['The Overthrow of Metaphysics through Logical Analysis of Language'] *Erkenntnis* 2 (1932).
Crowe, Michael J. *The Extra-Terrestrial Life Debate, 1750–1900*. Mineola, NY: Dover, [1986] 1999.
Darwin, Charles. *On the Origin of Species*. Edited by William Bynum. London: Penguin, [1859] 2009.
Darwin, Erasmus. *The Temple of Nature, or, the Origin of Society: A Poem with Philosophical Notes*. London: 1803.
Dembski, William A. "Schleiermacher's Metaphysical Critique of Miracles." *Scottish Journal of Theology* 49, 4 (1996): 443–465.
Dempsey, Michael T., ed. *Trinity and Election in Contemporary Theology*. Grand Rapids, MI: Eerdmans, 2011.
Derham, William. *Astro-theology: Or, A Demonstration of the Being and Attributes of God from a Survey of the Heavens*. London: 1715.
Descartes, René. *Le Monde*. (1667).
Dillenberger, John. *Protestant Thought and Natural Science*. Notre Dame: University of Notre Dame Press, 1960.
Dole, Andrew. *Schleiermacher on Religion and the Natural Order*. Oxford: Oxford University Press, 2010.
Frank, Manfred. "Metaphysical Foundations: A Look at Schleiermacher's Dialectic." Translated by Jacqueline Mariña and Christine Helmer. In *The Cambridge Companion to Friedrich Schleiermacher*. Edited by Jacqueline Mariña. Cambridge: Cambridge University Press, 2005.
Garber, Daniel. "Descartes and Occasionalism." In *Causation in Early Modern Philosophy: Cartesianism, Occasionlism, and Preestablished Harmony*. Edited by Steven Nadler. University Park, PA: Pennsylvania State University Press, 1993.

DOI 10.1515/9783110542301-008

–. *Leibniz: Body, Substance, Monad.* Oxford: Oxford University Press, 2009.
Gerrish, Brian. "Nature and the Theater of Redemption." In *Continuing the Reformation.* Chicago: University of Chicago Press, 1993.
Griffin, Michael. *Leibniz, God and Necessity.* Cambridge: Cambridge University Press, 2013.
Gould, Steven Jay. "Nonoverlapping Magisteria." *Natural History* 106 (1997): 16–22.
Gregory, Frederick. *Natural Science in Western History.* Boston: Houghton Mifflin, 2008.
Herschel, William. "On Nebulous Stars, Properly So-Called." *Philosophical Transactions of the Royal Society of London*, 1791.
–. "On the Nature and Construction of the Sun and Fixed Stars." *Philosophical Transactions of the Royal Society of London*, 1794.
Kant, Immanuel. *Universal Natural History and Theory of the Heavens.* Translated by Stanley Jaki. Edinburgh: Scottish Academic Press, [1755] 1981.
–. *Critique of Pure Reason.* Translated and edited by Marcus Weilgelt. Translated by Max Müller. London: Penguin, [1781, 1787] 2007.
–. *Critique of Judgment.* Translated by Werner S. Pluhar. Indianapolis: Hackett, [1790] 1987.
Käfer, Anne. "Kant, Schleiermacher und die Welt als Kunstwerk Gottes." *Zeitschrift für Theologie und Kirche* 101 (2004).
Keesee, Neal Kenneth. *The Divine Purpose in Schleiermacher's Doctrine of the Living God.* Dissertation. The University of Chicago, 1997.
Lamarck, Jean Baptiste. *Zoological Philosophy.* Translated by Hugh Elliot. London: Macmillan, [1809] 1914.
Lambert, J. H. *Cosmologische Briefe über die Einrichtung des Weltbaues.* Augsburg: 1761.
Lamm, Julia A. *The Living God: Schleiermacher's Theological Appropriation of Spinoza.* University Park, PA: Pennsylvania State University Press, 1996.
Laplace, Pierre-Simon. *The System of the World.* Translated by Henry Hickman Harte. Dublin: [1796] 1830.
Laudan, Larry. *Progress and Its Problems.* Berkeley: University of California Press, 1977.
Leftow, Brian. *God and Necessity* Oxford: Oxford University Press, 2012.
Leibniz, Gottfried Wilhelm. *Discourse on Metaphysics.* In *Philosophical Essays.* Edited and translated by Roger Ariew and Daniel Garber. Indianapolis: Hackett, [1686] 1989.
–. *On Nature Itself.* In *Philosophical Essays.* Edited and translated by Roger Ariew and Daniel Garber. Indianapolis: Hackett, [1698] 1989.
–. *Theodicy: Essays on the Goodness of God, the Freedom of Man, and the Origin of Evil.* Edited by Austin Farrer. Translated by E. M. Huggard. La Salle, IL: Open Court, [1951] 1985.
Leplin, Jarrett, ed. *Scientific Realism.* Berkeley: University of California Press, 1984.
Mariña, Jacqueline, ed. *The Cambridge Companion to Friedrich Schleiermacher.* Cambridge: Cambridge University Press, 2005.
–. *Transformation of the Self in the Thought of Friedrich Schleiermacher.* Oxford: Oxford University Press, 2008.
McCormack, Bruce L. "What Has Basel to Do with Berlin? Continuities in the Theologies of Barth and Schleiermacher." In *Orthodox and Modern: Studies in the Theology of Karl Barth.* Grand Rapids: Baker, 2008.
Nadler, Steven. "The Occasionalism of Louis de la Forge." In *Causation in Early Modern Philosophy: Cartesianism, Occasionlism, and Preestablished Harmony.* Edited by Steven Nadler. University Park, PA: Pennsylvania State University Press, 1993.

Newton, Isaac. *Opticks*. London: 1704.
Niebuhr, Richard R. *Schleiermacher on Christ and Religion*. New York: Charles Scribner's Sons, 1964.
Osiander, Andreas. *Ad lectorem de hypothesibus huius operis*. 1543.
Paley, William. *Natural Theology*. Edited by Matthew D. Eddy and David Knight. Oxford: Oxford University Press, 2008.
Pedersen, Daniel. "Eternal Life in Schleiermacher's *The Christian Faith*." *International Journal of Systematic Theology* 13, 3 (2011), 340–357.
–. "Schleiermacher and Reformed Scholastics on the Divine Attributes." *International Journal of Systematic Theology* 17, 4 (2015), 413–431.
Plantinga, Alvin. *Where the Conflict Really Lies: Science, Religion, and Naturalism*. Oxford: Oxford University Press, 2011.
Plato. *Timaeus*. Translated by Benjamin Jowett. New York: MacMillan, 1959.
–. *The Republic*. Translated by Desmond Lee. London: Penguin, [1987] 2007.
Popper, Karl. *The Logic of Scientific Discovery*. London: Routledge, [1932, trans. 1959] 2002.
Proudfoot, Wayne. *Religious Experience*. Berkeley: University of California Press, 1985.
Quine. W. V., "Two Dogmas of Empiricism." *The Philosophical Review* 60, 1 (1951), 20–43.
Rowan, Frederica, trans. *The Life of Schleiermacher as Unfolded in His Autobiography and Letters*. London: Smith, Elder, 1860.
Russell, Robert John. *Cosmology: From Alpha to Omega*. Minneapolis: Fortress, 2008.
–, Nancy Murphy, and Arthur Peacocke, eds. *Chaos and Complexity: Scientific Perspectives on Divine Action*. Vatican City: Vatican Observatory Publications, and Berkeley: Center for Theology and the Natural Sciences, 2000.
–, Philip Clayton, Kirk Wegter-McNelly, and John Polkinghorne, eds. *Quantum Mechanics: Scientific Perspectives on Divine Action*. Vatican City: Vatican Observatory Publications, and Berkeley: Center for Theology and the Natural Sciences, 2002.
Rutherford, Donald P. "Natures, Laws, and Miracles: The Roots of Leibniz's Critique of Occasionalism." In *Causation in Early Modern Philosophy*. University Park, PA: Pennsylvania State University Press, 1993.
Saunders, Nicholas. *Divine Action and Modern Science*. Cambridge: Cambridge University Press, 2002.
Schleiermacher, Friedrich. *Der christliche Glaube nach den Grundsätzen der evangelischen Kirche im Zusammenhange dargestellt*. Second edition. Edited by Rolf Schäfer. Berlin: Walter de Gruyter, 2008.
–. *Introductions to the Dialogues of Plato*. Translated by William Dobson. London: 1836.
–. *On the Glaubenslehre: Two Letters to Dr. Lücke*. Translated by James Duke and Francis Fiorenza. Atlanta: Scholars Press, 1981.
–. *Lectures on Philosophical Ethics*. Edited by Robert B. Louden. Translated by Louise Adey Huish. Cambridge: Cambridge University Press, 2002.
–. *The Christian Faith*. Edited by H.R. Mackintosh and J.S. Stewart, translated by D.M. Baillie, et al. Berkeley: Apocryphile, [1928] 2011.
–. *Brief Outline of Theology as a Field of Study*. Translated by Terrance Tice. Louisville: Westminster John Knox, [1811, 1830] 2011.
Schreiner, Susan. *The Theater of His Glory: Nature and the Natural Order in the Thought of John Calvin*. Grand Rapids: Baker, 1991.

Schwab, Johann Christian. “Prüfung der Kantischen Hypothese Ursprung des Ringes des Saturn, und die Berechnung der Achsendrehung dieses Planeten.” *Philosophisches Archiv* 1, 2 (1793): 1–21.
–. “Prüfung der Kantischen Hypothese von dem mechanischen Ursprung des Planetensystems.” *Philosophisches Archiv* 1, 2 (1792): 1–36.
Sherman, Robert. *The Shift to Modernity: Christ and the Doctrine of Creation in the Theologies of Schleiermacher and Barth.* New York: T&T Clark, 2005.
Spinoza, Benedict de. *A Spinoza Reader: The Ethics and Other Works.* Edited and translated by Edwin Curley. Princeton: Princeton University Press, 1994.
–. *Theological-Political Treatise.* Edited by Jonathan Israel. Translated by Michael Silverthorne and Jonathan Israel. Cambridge: Cambridge University Press, 2007.
Thomas Aquinas. *On the Eternity of the World.* Translated by Cyril Vollert. Milwaukee: Marquette University Press, [1964] 2003.
–. *Summa Theologiae.*
Torrance, Thomas F. *Divine and Contingent Order.* Edinburgh: T&T Clark, [1981] 1998.
Vogler, Candace. *Reasonably Vicious.* Cambridge, MA: Harvard University Press, 2002.
Ward, Keith. *God, Chance, and Necessity.* Oxford: Oneworld, 1996.
Wildman, Wesley. “The Divine Action Project, 1988–2003.” *Theology and Science* 2, 1 (2004): 31–75.
Wright, Thomas. *An Original Theory or New Hypothesis of the Universe.* London: 1750.

Index of Names

Index of Subjects

Bei Fragen zur Produktsicherheit wenden Sie sich bitte an:
If you have any questions regarding product safety,
please contact:

Walter de Gruyter GmbH
Genthiner Straße 13
10785 Berlin
productsafety@degruyterbrill.com